PAPERMAKING
FOR BASKETRY AND OTHER CRAFTS

EDITED BY LYNN STEARNS

Lark
Books

Front Cover: Intoxicated with Shadows of Flowers; Mary Merkel Hess; 18" x 10" x 10"; photo: Randall Tosh.

Back Cover: Untitled; Sue Smith; twined basket with mold-formed sheets.
      Morning Light Basket; Lissa Hunter; 10"d x 8-1/2"h.
      Square Basket; Marilyn Wold, handmade paper of ginger and Dracaena draco fiber; 10"h x 6" x 6".
      Untitled; Rosemary Gonzalez; paper vessel; photo: Carlos Quintanilla.
      Nest for Butterflies; Pam Barton.
      I'm Not Fooling This Time; Ellen Clague; 8"h x 12"w.

Production: Geri Camarda
             Sandra Montgomery
             Elaine Thompson

Library of Congress Cataloging-in-Publication Data

Papermaking for basketry and other crafts / edited by Lynn Stearns.

      p.   cm.

  Rev. ed. of: Papermaking for basketry. 1988.

  Includes bibliographical references (p.) and index.

  ISBN 0-937274-62-3

  1. Paper work. 2. Papermaking. 3. Basket making. I. Stearns,

Lynn. II. Papermaking for basketry.

TT870.P366 1992

676'.22--dc20   92-15002

                                         CIP

ISBN 0-937274-62-3

10 9 8 7 6 5 4 3 2 1

Published in 1992 by Lark Books
50 College Street
Asheville, North Carolina, U.S.A. 28801

First published in 1988 by
Press de LaPlantz, Inc., Bayside CA
under the title "Papermaking for Basketry"

Printed in the U.S.A. by
The Ovid-Bell Press, Inc.
Fulton MO

TABLE OF CONTENTS

INTRODUCTION	4
MEET THE ARTISTS	6
SUE SMITH	8
PLANT CHARTS	32
JUDY MULFORD	40
ALICE WAND	47
PAM BARTON	52
MARY LEE FULKERSON	63
ROSEMARY GONZALEZ	70
BETZ SALMONT	75
CAROLYN DAHL	79
MARILYN WOLD	91
HAWAIIAN PLANT CHARTS	105
SHARON BOCK	109
MARY MERKEL-HESS	119
LISSA HUNTER	124
GAMMY MILLER	140
ELLEN CLAGUE	144
KARRON NOTTINGHAM HALVERSON	152
DONNA RHAE MARDER	154
BIBLIOGRAPHY	156
SUPPLIERS	157
GLOSSARY	158
INDEX	160

Introduction

When we approached each artist to be in this book, we asked them to write about how they worked. We specifically asked them not to be democratic in their section, not to show the "right way," and not to include something because it's important to know. By showing so many options, we hope your creative spark plugs will ignite.

Creativity involves options—both finding and exploring them. As we are exposed to a new technique or medium, it is a time for questioning: For questioning why we work the way we do; why we are attracted to certain forms, lines, and designs; and why we start designing with the function, shape, or color in mind.

The book has been designed to work for you as a workshop and as an inspiration. The goal was to show different approaches taken by many artists, not to give one definitive view. There are as many ways to use paper in basketry as there are artists, and all are correct if they give the results that you want. This book highlights 17 artists, with each one sharing her own special approach to papermaking in basketry. Some take you step-by-step through the process, in the style of a workshop; others share their works and their inspiration, like a stroll through an exhibit. Although two of the artists presented here are not basket artists, their work bears a strong relationship to the rest of the work in the book, and we thought it would be fun to share them with you.

With this book you may need an openness to new definitions as well as new experiences. Are you limiting yourself by defining a basket through technique, material, or function? Do you have to make baskets? Could you make vessels, containers, forms, constructions, or art instead? What is the value of a definition anyway? It sets up parameters—parameters we can work within and develop a style within. Having everything open to exploration is too much. Exploration/creativity must be guided. But as we continue, we have the right to change our own parameters—to redefine how we work.

Cutting It Out

I (Shereen LaPlantz), have always been a bit jealous of my husband. As a jeweler, when he doesn't like some part he is working on, he just cuts it out. He also keeps a "parts box" of little pieces that can be added onto something else. Jewelers can just solder a new part on—anywhere. It has never seemed fair. David can add and subtract from an art piece however he wants. Yet when I need to subtract, I have to rip the whole thing apart and start over. With paper, it seems that additions and subtractions are possible. I can imagine cutting a piece of paper. And I can imagine casting another sheet onto existing paper. This could lead to a whole new way of thinking for basket artists!

Tips On The Instructions

Before this book went to the printer, we proofed the instructions by using them. It was fun! We have a vase full of frosted teasels and several pieces of interesting paper. We had no trouble following the instructions, but we do have some tips to pass on.

Tip: Listen when you are told how to cut up the plant materials. When one of us harvested and cooked up a batch to make sheet paper, it wouldn't blend and just continued to look like a messy bunch of sticks. So we called Marilyn Wold, who said we hadn't cut the sticks short enough. Not to worry, they could still be cut at this stage. So we cut and attacked with vigor. The blending went well, as did the sheet forming. But the sheets disintegrated on touch. Back to Marilyn, who said we had cut the sticks too short!

Tip: Use the correct materials. Another one of us made some lovely pastel frosted teasels with food coloring as a dye, and the teasels soon faded to white. Tip: Use your meat tenderizer on a wooden or other giving surface. We made a tapa beater that works the same as a meat tenderizer. One of us used the tapa while I used an aluminum meat tenderizer. I worked on the concrete outside my house, and the aluminum spikes blunted immediately. They worked much better against a wooden plank.

Tip: When I've watched others couching a paper sheet, I've noticed they lay the mold down with regular motions, but lift it off with a swift little jerk. I could get the sheets off that way, or by sponging so thoroughly that the sheet started transferring during the sponging process.

Tip: Be careful of the chemicals, toxic plants, etc. You only get one life, so err on the safe side. This book is for artists experienced in the studio, who are aware of proper safety precautions when working with chemicals, dyes, etc. Papermaking uses some chemicals that can be harmful if not properly used. You should always practice common sense and good judgement: keep chemicals and equipment separate from food and eating areas; always clean your work area thoroughly when you are finished working for the day; wear rubber gloves and a mask; and read and follow instructions on the chemicals you use. The artists have suggested the safety tips that they personally use. Some of them do work in the kitchen, which is not recommended, but if you must, do not use kitchen equipment that will also be used for preparing food. When using poisonous chemicals or plants, or combinations that could cause toxic fumes, don't let them splash onto kitchen counters. Make sure you have good ventilation. Remember that even non-toxic plants or chemicals can cause allergic reactions in some people, and some plants that are not poisonous can become toxic when combined and cooked. Even those pulps containing edible materials should be handled with care. And always remember to turn off and unplug beaters or blenders before putting your fingers in to check the consistency of the pulp.

Please remember as you go about collecting materials to be extremely cautious about endangered species, both plants and animals.

Miscellaneous

Occasionally, trade names have been used in this book. They have all been capitalized. Unless otherwise indicated, all photographs are by the artist.

Have Fun

Each of the authors included the phrase "Have fun." Do! This is the best part of a new experience. Play until the new medium becomes your own.

SHEREEN LAPLANTZ and LYNN STEARNS

■ Meet The Artists

Pam Barton, of Volcano, Hawaii, received a BFA degree at the University of Hawaii and also studied at the Honolulu Academy of Arts. She has been exhibiting her artwork in the state of Hawaii for over 20 years, and her fiber works have been shown locally, nationally, and internationally in exhibitions. Barton teaches workshops and classes in basketry and papermaking.

Sharon Bock of Bonner Springs, Kansas, received a BFA degree from Washburn University and studied three-dimensional handmade paper at the Kansas City Art Institute. She is a recipient of a fellowship from the Mid America Arts Alliance/National Endowment for the Arts, and her work has appeared in numerous juried and invitational exhibitions throughout the United States. She conducts workshops on sculptural handmade paper.

Ellen Clague, of Bethel, Connecticut, maintains a fiber studio, teaches workshops, and exhibits her work nationally in juried and invitational shows. She attended Denison University in Ohio, and also studied painting in Germany. She worked as gallery coordinator at the Brookfield Craft Center in Brookfield, and assistant manager at the SONO Craft Complex in Norwalk.

Carolyn Dahl, of Houston, Texas, works in dyed and handmade paper vessels and handpainted silk wall hangings. Her work has been published in many magazines, including *American Craft, Fiberarts,* and *The News Basket.* In addition to producing work for galleries and exhibitions, Dahl lectures and conducts workshops.

Mary Lee Fulkerson lives in Palomino Valley, Nevada, near the Pyramid Lake Indian Reservation. The proximity to the reservation has had a strong influence on her work. She has written for publications and teaches papermaking.

Rosemary Gonzalez, of Austin, Texas, studied at Pan American University, University of Houston, and the Museum of Fine Art in Houston. Gonzalez was a commercial artist, then an art teacher, and now devotes her time to the development and marketing of paper pottery. She is represented by several galleries and two world trade centers.

Karron Nottingham Halverson, of Hilo, Hawaii, is a papermaker who works with mixed media constructions. She received a BS degree from the University of Wisconsin, and also attended the University of Minnesota-Tlaxico in Oaxaca, Mexico. She is represented by galleries in Hawaii and the mainland, and her work has appeared in exhibitions throughout the United States. She teaches and conducts workshops.

Mary Merkel-Hess lives in Iowa City, Iowa, where she attended the University of Iowa and graduated with an MA and an MFA. She received her bachelor's degrees at the University of Wisconsin and Marquette University. She has exhibited her vessels in the United States and Canada. Her work can also be seen in *The Basketmaker's Art,* and *The News Basket.*

Lissa Hunter, of Portland Maine, exhibits in galleries around the country. She works in her studio and teaches workshops. She received an MFA degree in Textile Design from Indiana University. Her work is represented in *The Basketmaker's Art,* and *The News Basket.*

Donna Rhae Marder, of Chicago, Illinois, has a BFA in fine arts from the University of Chicago, and has also attended the Art Institute of Chicago and the Hyde Park Art Center in Chicago. Her work has appeared in exhibitions around the country, as well as in Toronto.

Gammy Miller lives and works in New York City with her family. She has worked with fiber since 1972, first with knotted neckpieces and later with coiled basketry. The continuum throughout her work has been the influence of the sea, where she and her family spend their summers.

Judy Mulford is a California basket artist, teacher, and lecturer who has done textile research throughout Micronesia and in Mexico. She has exhibited in shows around the United States, Canada, and Africa, and her pieces are in numerous collections. She is the author of *Basic Pine Needle Baskerty.* Mulford says that she owes all of her papermaking knowledge to her dear friend and teacher, Sue Smith.

Betz Salmont, of Manhattan Beach, California, was a papier-mâché artist, sculptor, and weaver before becoming a basket artist. As a recipient of a California Arts Council Grant, she teaches basketry throughout the state. She has taught at the Craft and Folk Art Museum and at basketry symposiums. Salmont has exhibited her baskets in the United States and Africa.

Sue Smith, of Fort Worth, Texas, is a basket artist, teacher, writer, and conference coordinator for Press de LaPlantz. Currently her basketry focus is the exploration of handmade papers for her sculptural forms. Her work has been shown in numerous juried and invitational exhibitions, and she has taught extensively throughout the United States. She is also the author of *Natural Fiber Basketry.*

Lynn Stearns, of Yorkville, Illinois, graduated from Northern Illinois University with a BFA degree. Her basket art has been on exhibit in the United States, Canada, and Africa.

Alice Wand, of Saranac Lake, New York, graduated from the University of Wisconsin at Milwaukee with a BFA degree. Her work has been exhibited throughout the United States, and she is currently represented by galleries all over the country. Wand has been featured in articles in *The News Basket* and in *Fiberarts.*

Marilyn Wold, of Kailua-Kona, Hawaii, studied art at the Oregon School of Arts and Crafts and the Portland Museum Art School. She has been working in paper since 1978 when she studied papermaking with Lillian Bell. Wold teaches workshops in papermaking, and her work has appeared in exhibitions in Hawaii, Oregon, and Japan.

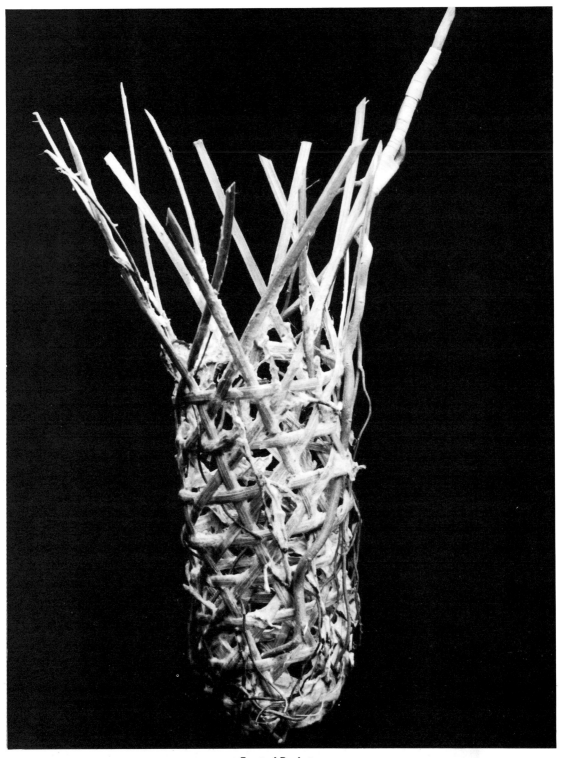

Frosted Basket.

SUE SMITH

Basketry–
An Adventure with Paper

One of the many delights of basketmaking is the search for interesting materials. This search has led to the exploration of the many fibers available. Plant fibers are my favorite, and I enjoy harvesting them to create contemporary basket forms.

Plant fibers are also materials for papermaking. Paper is made by the layering of plant fibers, broken down into cellulose and suspended in water, then lifted from the water on a screen frame to form sheets.

As I make my baskets, lovely bits of materials fall to the floor around me, bits of plant material that could be used to make paper. What are the possibilities of blending these fibers into paper pulp and achieving a fusion of paper and basketry? This question generated an investigation of raw materials suitable for paper forming, as well as basketry.

One of my first efforts, and still one of my favorites, is "sweep up paper." "Sweep up paper" is made from the extracted cellulose fiber from plant materials left over from a current basket project, plus odds and ends from my herb and flower beds. This mixture makes an intriguing and highly textural sheet of paper. At first, I didn't know if this mixture would hold together as paper, so I would add a small amount of recycled paper pulp or abaca (a ready-to-use pulp) for strength. Continued experimentation revealed that many of these plant fibers had not only luster and texture, but also sufficient strength of their own to form quality paper.

Untitled; coiled basket of reed, raffia, and "sweep up paper"; 3-1/2" h x 5" d.

Through this experimentation in making the paper myself, I feel a partnership with the materials. My rapport with the fibers always grows during the process of forming them into the basket. My goal has been to create a sense of harmony between the fibers on the surface and those in the structure—a unified design. As I construct the form of my basket, the smell and feel of the materials are pleasing and exciting to me. As these fibers flow through my fingers, a creative momentum evolves, making me and my baskets richer for the experience. There is a wealth of material suitable for paper forming, ranging from plant materials to recycling paper, and even ready-to-use pulp. Each material has its own special characteristics. Detecting the spirit of each individual material, as well as blends of materials, stimulates endless possibilities of transforming these raw materials into a collaboration with basketry. This sensitivity to the materials is the foundation for my personal statement in this medium.

Any plant may be considered for investigation because paper is primarily made from cellulose, and all plants have some cellulose content. Many plants have been tested for their papermaking qualities. An excellent resource book is *Plant Fibers for Papermaking* by Lillian Bell. Generally, suitable plants for papermaking have been divided into four fiber categories: bast, leaf, grass, and seed.

Bast fiber is the phloem, the inner bark or core in branches and stems. This fiber is usually long and strong, making it ideal for papermaking. Suggested plants include birch, wisteria, mulberry, fig, willow, milkweed, kudzu, hollyhock, and osage orange.

Leaf fibers are ideally chosen from plants with long, sword-like leaves. Examples include iris, gladioli, yucca, agave, and New Zealand flax. Leaves from trees and shrubs are generally not strong enough by themselves, but can be added to a stronger fiber such as abaca.

Untitled; coiled basket of Boston ivy and "sweep up paper"; 6" h x 7" w.

Grass fibers tend to produce brittle paper for sheets, although they make interesting textured paper when mixed with a stronger fiber base. They also work well into the weave of the basket. There are many types of grasses available. Some suggested for papermaking include wheat, rice straw, bulrushes, cattails, and papyrus.

Seed fibers can be used, but seem to be very difficult to process by hand.

Primarily, I use the same plant materials as I do in my baskets and dyes. As I am gathering materials for my baskets, working in my garden, and even when preparing a salad, I consider the possibilities of using these plants in papermaking. Most of my harvesting is done after the plant has reached maturity, flowered, and gone to seed, generally late summer to fall. I am conservative in gathering, leaving more than I take so that the plant can continue a hardy existence. Another opportune time to gather materials is when contractors are clearing trees, shrubs, vines, and grasses.

My search for basketry and papermaking fibers has been a continuous one. I keep a journal which includes the name of the plant, a pressed sample, where and when I gathered it, and how I prepared and used it. After working with a material, I like to keep it for a period of time to see how it will hold up. Historical plant research and plant recommendations have aided my selections, although I have not found an easy way to determine a specific plant's usability until I have worked through the whole process.

During my experiments, I have learned a great deal about plants; but perhaps more important, I have realized how much more there is to learn. The plants that are included in the chart at the end of this chapter are only a sampling of the wide selection available. Using the plants abundant in your area will provide much interesting experimentation and will broaden your perspective. Experimenting and sharing your results will lead you, as it did me, to a fuller appreciation of the world around us.

Tools and Equipment

Most of my papermaking is done in the kitchen and on the patio. Many of the tools serve a dual purpose in papermaking and basketry. To make your own paper you will need:

- mold and deckle (make your own or purchase from supplier)
- fibers (gather your own plant materials, order ready-to-use pulp and unbeaten fibers from suppliers, or recycle papers)
- vat (plastic dishpan, busboy's tub, or a plastic laundry sink on legs)
- couching cloth (Pellon, sheeting, cotton diapers, or Handi Wipes)
- blender
- stainless steel meat tenderizer or wooden mallet
- rubber gloves
- towels
- sponges
- large cooking pot (stainless steel or enamel)
- measuring spoons and cups
- large spoon
- sodium carbonate (washing soda and soda ash are other names for this chemical)
- TSP (trisodium phosphate is a mild alkali which can be substituted for sodium carbonate)

- strainer (nylon net bag, panty hose, colander, or wire mesh strainer)
- storage containers (gallon size plastic jars or Zip Lock plastic bags)
- sharp knife, clippers, and vegetable scraper
- petroleum jelly or non-stick aerosol spray
- dyes (fiber reactive or natural)
- methyl cellulose
- cheese cloth
- acrylic medium
- varnish
- potpourri, etc., according to the needs of your imagination

In the papermaking process, there are precautions to keep in mind and practice:

- *Keep your papermaking tools and equipment separate from your cooking equipment.*
- *NEVER leave a pot cooking or blender running unattended.*
- *Wear rubber gloves when using alkalies, dyes, and sizing.*
- *Wear a dust mask when handling a lot of dry fibers.*
- *Do not use a cooking pot larger than your burner.*
- *Work in a well-ventilated area.*

Preparing the Plant Fibers

Once the plants have been gathered, they must be cooked. Cooking removes pectin, starch, lignin, and other non-cellulose materials which can adversely affect the strength of the paper.

In preparing the various plant materials for cooking, I have tried a variety of methods, hoping to hasten this least favorite part of the process. For example, I used a leaf shredder on small branches, trying to obtain bast fiber. After a great deal of noise, dust, and sneezing, I decided I preferred to strip the bark by hand with a knife or vegetable peeler.

Bark is easier to peel when the branch is freshly cut. This is especially successful in the spring when the sap is up and there is a high moisture content in the bark. Using a vegetable peeler, peel the outer bark in thin strips from the butt to the tip. The bark can also be scraped off with a knife. To obtain the bast fiber, cut a slit with a knife into the butt end of the branch down to the core and peel down to the tip. If it does not peel easily, the branch will need to be steamed.

To steam, cut branches to fit your cooking pot. Fill the pot halfway with cuttings and cover with water. Bring to a boil and simmer, covered, over medium heat for about 30 minutes. The branches must be peeled immediately after steaming. The bast fibers can now be cooked for pulp or frozen for later use. If frozen, there is no need to thaw before cooking.

Leaf fibers with a tough water-repellent outer layer need to be scraped before cooking. These plants include yucca, sansevieria, pineapple, agave, and canna lily. The leaves can also be pounded with a meat tenderizer or wooden mallet to obtain the long thread-like strands. Other leaves and grasses can be chopped in a food processor, leaf shredder, or cut quietly with scissors or a knife.

After preparing the plant materials, I usually cover them with water and soak for 24 hours before cooking. I think this helps to clean and hydrate the fibers.

Preparing the Plant Fibers

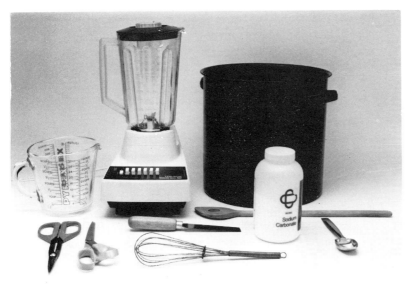

Tools and equipment to be used in the pulp-making process: cookpot for pulp, blender, measuring cup, scissors, and spoons.

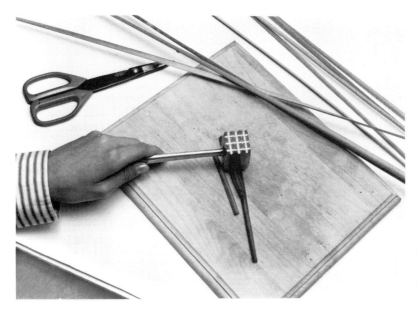

Tough stems need to be crushed with a meat tenderizer or wooden mallet.

Cut plant materials into short pieces.

Cooking the Fibers

Because of the many variables in working with plant materials, it is difficult to give specific cooking directions. Generally, I fill the cooking pot about halfway with the soaked plants and cover with water. Initially I used lye, but decided there were too many negatives, such as the extremely caustic and poisonous fumes released during the cooking process. Now I use a milder alkaline substance, in the ratio of one tablespoon to each quart of water. Two mild alkalis are TSP and sodium carbonate. Soda ash and washing soda are considered the same chemical as sodium carbonate. (Since Arm & Hammer has changed their washing soda formula, I no longer recommend it for papermaking.)

This alkaline solution reduces the fleshy part of the plant to usable pulp more rapidly. Bring this mixture of plant materials, alkali, and water to a boil and simmer approximately two hours. Check the progress of the material by lifting a small amount of the fiber and noting its appearance. When sufficiently cooked, the plants should be slippery and mushy, with the fiber broken down into a stringy substance. Very tough fibers, such as cedar, may require longer cooking time. An alternative would be a second cooking with fresh water and an alkaline substance.

After the cooking is completed, the fibers must be thoroughly rinsed to remove all non-cellulose material and alkaline substance. Pour the mixture into a colander or strainer and let water run slowly over the fibers while gently stirring. Cleaned fibers are ready to be beaten or stored.

The cooking process I use for dye plants such as osage orange, sage, and yellow clover differs slightly. The dye plants are simmered in clear water for one hour. After simmering, let the dye cool with the plants in it. When cooled, strain out the plant fiber into a colander or strainer. Save the dye in a container in the refrigerator or freezer. Replace the plant fibers in the cooking pot and proceed with general cooking procedures. I later add the dye to the vat in the paper-forming process.

My information on papermaking is based on personal experience that is an on-going learning process. In the beginning I pursued exploration of different materials and integrated them into my baskets, without a great regard for purity. As my work progressed, the archival quality of the paper became a consideration. I am striving to develop a quality of paper that is not only beautiful but long lasting, with integrity in my art forms. In pursuit of this goal, and through further research, I have learned that certain chemicals are detrimental to cellulose and may cause the paper to become brittle and yellow. For example, mordants are commonly added to the dye bath. Mordants are chemicals used to bond the dye to the fiber. These mordants—alum, chrome, copper, tin, and iron—vary from slightly to deadly poisonous. I used to mordant the dye for my pulp with alum (one of the safest chemicals in this group), but now I understand that it is an acid that will destroy the paper, and consequently I don't use it. Instead, a retention agent to help with the bonding has been recommended by Elaine Koretsky of The Carriage House Handmade Paper Works. Chlorine laundry bleach is another substance that I no longer use. (When the decomposed plant fibers had a distinct odor, I would add a teaspoon of bleach to eliminate the odor and to lighten the color.) Hydrogen peroxide now seems a safer solution.

To store the drained and cooled fibers, I put them in freezer bags or cartons, cover with water, and freeze until ready to use.

Cooking Plant Fibers

Soak plant materials for 24 hours.

Periodically stir the pulp.

Drop a small handful of pulp into the blender.

Beating the Fibers

Beating softens and hydrates the fibers, which facilitates the bonding process. Beating times will vary, not only with each plant, but also with the type of paper desired. Fiber beaten for a short time—15 seconds—will produce a soft, blotter-like paper with strands or pieces of the fiber evident. Well-beaten fiber has more bonds and makes a strong, fine paper that is often used for calligraphy and book paper.

There are several methods of beating the fibers: you can hand beat with a wooden mallet, or let a machine do the beating with a Hollander, a paint stirrer on an electric drill, or a kitchen blender. I primarily use my kitchen blender, which does an adequate job of separating the fibers.

Cut the fibers into 1/2-inch lengths. Place a small handful of fibers (approximately 1/4 cup) into the blender. Add water and cover with the lid. The blender should never be more than 3/4 full, using 1/3 pulp and 2/3 warm water. Always begin with the slowest speed. Watch and listen closely. Some fibers may become entwined in the space under the blade and burn out the motor. Beat 15 seconds. Stop the blender and examine the fibers suspended in the water. You want the pulp to be well blended and free of fiber clumps. Continue beating and checking every 15 seconds until desired texture is obtained. When the fiber and water are smoothly and evenly blended, the process is complete. Note: always unplug the blender before putting your fingers in to check consistency.

If commercially available linters or recycled paper are used, cut them into 1/2-inch pieces and let them soak overnight in water. Construction paper, kraft paper, stationery, blotter paper, and other decorative papers are all suitable. After the paper bits are soaked, follow the procedure above.

At this point, dyes, buffering agents, and sizing can be added to the pulp. Dyes are soluble chemicals that penetrate the surface of the fiber and color it. There are three types of dyes suitable for papermaking: direct, fiber-reactive, and natural. I have used fiber-reactive and natural dyes successfully. For using color in papermaking, an excellent reference book is *Color for the Hand Papermaker* by Elaine Koretsky.

Buffering agents are substances which are added to the pulp to counteract the effects of acidic chemicals and pollutants in the atmosphere. Calcium carbonate and magnesium carbonate are two base chemicals recommended as protective buffers.

Sizing decreases the absorbency of the paper and also gives it additional strength. Paper can be sized in two ways: either internally in the pulp in the blender; or externally, after the sheet is formed and dried. Unflavored gelatin provides an easy, light internal sizing. Just blend one small envelope, prepared according to package directions, to a quart of pulp. Beat this mixture immediately and thoroughly into the pulp in the blender. Be sure to use warm water. Internal sizing can be purchased in liquid form from commercial papermaking suppliers. Methyl cellulose is a versatile medium which can be added to paper pulp to promote bonding and give added strength for paper casting. It is also an excellent non-acidic, non-staining adhesive.

External sizing can be done after the sheets have been formed and allowed to cure at least five days to two weeks. The paper should be thoroughly dry. Size by spraying with a fabric protector, Scotchgard, or a silicone spray. Acrylic matte finish spray is also satisfactory. I recommend that you use sizing to protect your work.

Continue making pulp until you have approximately a gallon. I usually pour off excess water from the blender before pouring the pulp into the gallon container to obtain a concentrated form of pulp. The pulp is now ready to form into paper.

Processing the Pulp

Beginning with the lowest speed, blend the pulp.

Drain excess water from the pulp before storing.

Equipment used in forming the pulp into paper: vat, mold and deckle, strainer, sponges, felts, and materials for creating textures.

Forming the Paper

The only equipment you will need during the paper-forming process is a mold and deckle, a vat, and felts or couching sheets. The mold is a wooden frame covered with a screen on which the paper is formed. The deckle is a second wooden frame which is placed on the mold to contain the pulp on the screen during the formation process.

The mold and deckle can be purchased from several sources listed in the appendix under Suppliers. However, a simple, inexpensive mold and deckle can be constructed.

Supplies and tools needed to build a simple frame:

- wood strips (1" x 2" or 1" x 1") cut to desired lengths
  (use 1" x 2" for 12" square or larger frames)
- fiberglass window screening material
- four 1-1/2" x 3/8" metal corner angles if you are using the
  1" x 2" wood strips
- small screws (to secure the corner angles)
- small finishing nails (to nail together 1" x 1" framing material)
- heavy-duty staple gun (not a desk stapler)
- duct or gaffer tape

If you are making a frame for small sheets of paper, up to about 8" x 10", use the 1" x 1" wood. Place the longer strips (10") between the ends of the shorter piece and secure with waterproof glue and small nails. If your frame is to be larger than 8" x 10", I recommend using 1" x 2" wood for strength. Use waterproof glue to secure the corners, then add one corner bracket to each corner with small wood screws.

Next, cut the fiberglass screen material just short of the outside edges of the wooden frame. Secure the screen to the frame (brackets on bottom, if they are used) using a heavy-duty staple gun. Alternate stapling from side to opposite side in order to stretch the screen evenly.

Once the screen is securely attached and taut, use strips of duct tape to cover the edges of the wood. (The tape may need to be cut lengthwise to fit.)

A frame larger than 11" x 14" may need additional support to keep the screen from sagging when it is covered with pulp. This can be done by adding hardware cloth, a galvanized wire screen with a half-inch mesh. First secure this to the frame with a stapler; then put the screen material on top of it and secure with staples.

An embroidery hoop can be used as a frame to obtain a variety of sizes and shapes of sheet paper. By simply cutting the window screen slightly larger than the diameter and pressing the second hoop into place, you have a simple, inexpensive round form. No stapling or gluing is necessary. The tension of the screen can be adjusted by removing the screen and pressing the hoop back together. (The hoops will warp with prolonged usage.)

The vat is a deep tray or tub, large enough to accommodate the mold and deckle with the pulp. I found the busboy tubs used in restaurants to be ideal for use with small molds. Felts are any material placed between the wet sheets of paper during the couching process. The felts separate the sheets in a stack or post, and aid in the drying. These can be Pellon, sheets, printmaking felt, cotton diapers, or Handi Wipes. The surface of the felt is a determining factor in the texture of the paper's surface. Textured materials such as burlap, towels, bark, and sections of chair caning patterns add impressions to the surface of the paper. The felts should be wet before couching. Wet felts will prevent the fibers from tearing during the transfer from the screen.

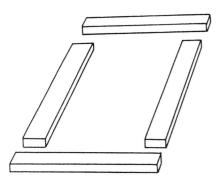

Place longer strips of 1" x 1" wood between the ends of the shorter pieces.

A simple frame can be made from an embroidery hoop.

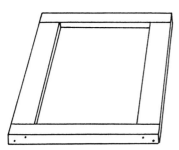

Secure with waterproof glue and small nails.

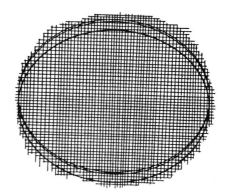

Cut the window screen slightly larger than the diameter of the hoop and press the second hoop into place.

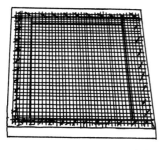

Cut the fiberglass screen materials just short of the outside edges of the wooden frames.

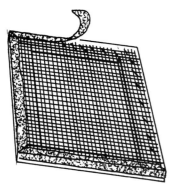

Once the screen is securely attached and taut, cover the edges of the wood with duct tape.

You now have an inexpensive round form.

Fill the vat with slurry (a blend of pulp and water) to the point where the mold and deckle can be completely immersed. The proportion of concentrated pulp to water depends on the desired thickness of the paper. The thicker the consistency of the pulp to the water, the thicker, more blotter-like the paper will be. If I am filling the vat directly from the blender, it takes about 10 blenders full. Using concentrated pulp, my proportions are generally six blenders of pulp to four blenders of water for a busboy tub. Experimentation will determine what you like best.

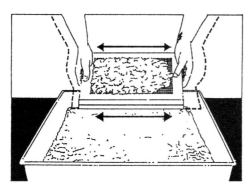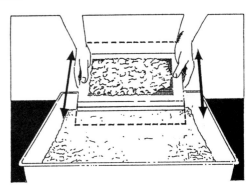

Two views of Vatman's Stroke.

Stir the pulp thoroughly so that the fibers will be suspended in the water. Grasp the mold and deckle together securely and start at the far end of the vat with the mold and deckle perpendicular to the pulp. Lower the mold and deckle into the vat in one continuous motion, leveling them out horizontally under the suspended pulp. Lift the mold until just clear of the pulp and then gently shake sideways, backwards, and forwards. This is called "the vatman's stroke" or "throwing off the wave." This procedure will help to even out the pulp and interlace the fibers.

Before removing the deckle, tip the mold and deckle at a 45 degree angle to drain for a few seconds. Remove the deckle and examine the paper. If it meets your standards, you are ready to dry it. If it is unacceptable, the pulp can be returned to the vat by turning the screen upside down on the water. Do not roll or scrape the pulp from the screen into the vat because it will make lumps in the pulp mixture.

Untitled; twined basket of multiple mold-formed sheets of handmade paper and dyed reed.

Forming the Paper

Lift the screens out of the vat. The deckle will hold the pulp on top of the screen.

The newly formed sheet of paper can be couched onto a textured surface or a damp felt.

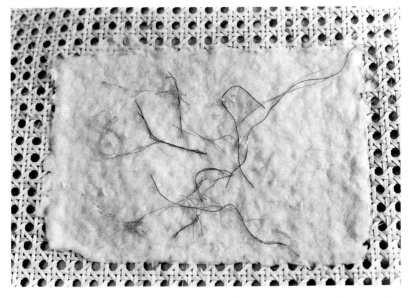

Threads can be sandwiched between thin sheets in the layering technique.

Drying the Sheet

The first step in drying is to transfer the paper from the mold to the felt. This step is called couching, from the French word which means to lie between the blankets. The couching, or transfer process, is easier if the felt or couching cloth is damp. Couching can be a very wet, sloppy process. This can be controlled by placing the damp couching cloth on top of a folded beach towel or old army blanket. Smooth the couching cloth. Grasp the mold on the longest side. With a rolling motion, turn the screen, pulp side down, onto the felt, and press down on the screen with a damp sponge to transfer the new sheet from the mold to the couching cloth. Lift the mold carefully and the paper will stick to the damp cloth. To form additional sheets, a post (stack) of paper may be made by placing another damp couching cloth directly on top of the newly made paper and repeating the dipping and couching process.

After dipping and forming several sheets, check the consistency of the pulp and water in the tub. There will be fewer fibers in the same amount of water. Add blended concentrated pulp as needed. When the desired number of sheets have been formed, a rolling pin may be used to gently squeeze out excess water. If you have a book binding or flower press, place the post on the bottom board. Carefully position the top board and tighten screws. Depending on the climate and the weather, you may leave the press set up overnight to dry. Papers may mildew if left in the press for more than a day.

An alternative to using a press is to dry your papers with an iron. Keep the newly formed sheet between two cotton sheets and press it with a dry, warm iron. Keep the iron moving and turn the sheets frequently. A delightful way to watch your newly formed paper dry is to couch it directly on the window. This produces flat paper with one very shiny side.

Paper can be air dried by placing it on racks. Let the newly formed paper remain on the couching cloth for about an hour before separating. Then lightly press four to five sheets together and return them to the racks. This allows the papers to dry slowly, so they won't warp or wrinkle. They can be pulled apart after they dry.

Untitled;
double woven
basket of willow
and willow paper;
11"h x 8"d.

Applying Pulp to Basket

Swirl the pulp to disperse the fibers into the desired areas of the basket.

Use a sponge to press the fibers into the weave and to absorb water.

Layers of assorted papers can be applied to the damp surface.

Exploring Variations in Paper Forming

Because of the tremendous potential of paper and the wide range of plants, the artist has great control and flexibility with variations in the paper-forming process. After learning the basic sheet formation, explore these possibilities:

- Vary the shape and size of the screen and/or deckle. Try cookie cutters as deckles.
- Vary the pulp by adding a potpourri of petals and leaves to the vat before the sheet is formed.
- Laminate two sheets of different colors, textures, and/or shapes together. Newly formed paper will adhere to paper when it is damp, so you can couch two sheets together.
- Embed objects such as feathers, bits of thread, leaves, and lace while the paper is still wet on the mold, and then couch it onto the felt.
- Sandwich thin objects such as fern, butterflies, and pine needles between two very thin, newly formed sheets.
- Couch the newly formed sheet over a textured surface such as bark, plastic meat trays, or sections of chair caning patterns for a bas relief.
- Manipulate the surface of the paper while it is on the screen by spraying with a plant mister.
- Build up forms with pulp applied to the sheet by squeezing the pulp from a turkey baster.
- Roll the newly formed sheets into coils, much like clay.
- Think of the characteristics of paper: strong/fragile, opaque/translucent, thick/thin, smooth/rough. It can be molded, folded, crumpled, cut, torn, punched, pierced, and embossed. With all of this in mind, exploration is an ongoing process.

When I began working in this exciting new medium, my goal was to explore ways that handmade paper could enhance and/or interact with my basketry forms to become an integral part of the design. As I experimented with the process and the different fibers, I was overwhelmed with the possibilities of making paper out of so many things. Paper is a wonderful material that can be an end unto itself. I had to focus on my goal to integrate this medium with basketry.

My first approach was to use sheets of my handmade paper. To do this, lay the sheet over the surface of the basket and tear it to fit the desired shape. Torn edges are softer and tend to blend more effectively than cut edges. Next, dip the torn paper into a methyl cellulose and water solution. Eliminate excess adhesive and press the paper into the weave of the basket. Repeat the process, overlapping layers of paper to create the desired surface texture and color. When the paper is completely dry, paint the surface with an acrylic medium gel to seal it.

Another method I use is to cast pulp into a form. This cast paper form becomes the foundation that the basket is woven into and around. Some of the forms that I have

Applying Pulp to Basket

This twined basket was made from willow, Dracaena, blue stem grass, and handmade paper of willow and millet; 7" h x 14" d.

Side view of twined willow basket.

Paper pulp can be cast into a wide assortment of molds.

used include shells, bones, clay containers, and plaster molds of faces, hands, and vegetables. After selecting the object, spray the surface with a non-stick aerosol spray or lightly coat the surface with petroleum jelly for easy release of the object from the paper after it dries.

Add methyl cellulose to the pulp during the beating process for added strength during the casting process. A tablespoon of this solution to a pound of pulp has worked effectively. Too much methyl cellulose will adhere the paper cast to the mold.

Slowly pour the pulp into the form, allowing it to flow into the contours and crevices of the mold, about one-half inch deep. After the paper has begun to set, use a sponge to press the pulp into the contours and to absorb excess moisture. For extra strength, reinforce with a layer of cheesecloth. Cheesecloth may be soaked with methyl cellulose or thinned white glue before layering it over the pulp in the mold. Cover with another layer of pulp. Let the pulp dry thoroughly before removing it from the mold.

The basket may be constructed first, and the cast paper forms then incorporated into the shape of the basket. To secure these forms, slowly and carefully pour pulp, with methyl cellulose adhesive, around them in the basket.

My favorite method of incorporating paper into my basket is "frosting." To frost your basket, dip the basket, instead of the mold and deckle, into the vat of pulp. Following a procedure similar to forming sheets, dip the basket down into the vat under the slurry. Bringing it up horizontally, gently swirl the pulp to disperse the fibers into the desired area. Press these fibers into the weave of the basket with a damp sponge. The proportion of pulp to water in the vat will determine the amount of frosting that clings to your basket. This procedure interacts splendidly with the random weave technique.

Summer Shadows; twined basket of handmade papers, pine needles, and reed;
17"w x 13"h.

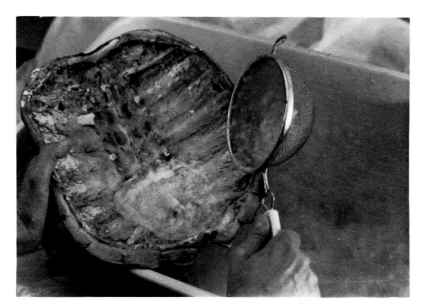

Molded Paper

Using a turtle shell as the mold, I pour the pulp into a strainer and then into the mold.

Gently press the pulp into the crevices of the mold. Tilt the mold to drain while working.

Paper cast from the turtle shell.

Incorporating Paper with Basket

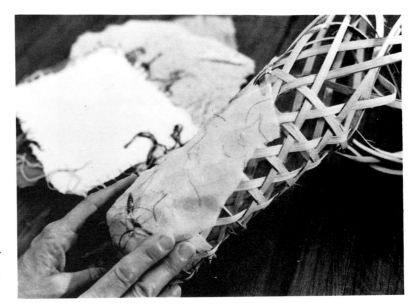

Press the sheet of paper gently into the weave of the basket.

Tortoise Shell; plaited basket of flat reed, Boston ivy, redwood roots, and the paper cast of the turtle shell; 14" l x 12" w x 9-1/2" h.

Back view of Tortoise Shell.

Untitled twined basket of multiple mold-formed sheets and reed.

Plaited basket of birch, cedar, and handmade paper; 16" h x 5" d (including branch).

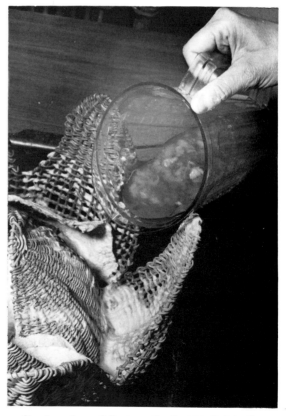

Slowly and carefully, pour the liquid pulp into the basket to seal the mold-formed shapes in place.

As I continue to work with this exciting medium, different fibers lure me on to further exploration into the interaction of paper and basketry. The inherent textural characteristics of handmade paper was the primary focus of my collaboration of paper with basketry. Experimenting with different techniques expanded my surface embellishment. Nature has always been an inspiration for me, as well as a source of materials. Birds' nests with delicate bits of paper entwined in them caught my attention. Wasps—nature's own papermaker—with their beautifully shaped nests of layers of extraordinary paper, also caught my eye. My investigation of papermaking in basketry reflects these themes from nature.

Forms from nature: oriole nest, white-eyed vireo nest, basket.

Hornet's nest.

All photos by David Smith.

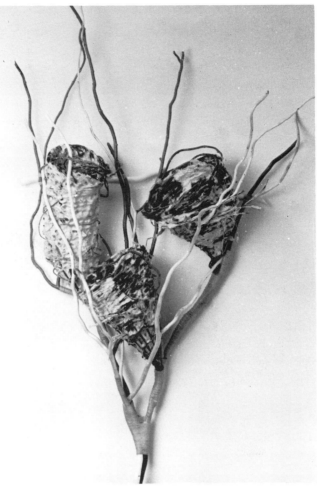

Nests of Willow; twined baskets of willow
frosted with willow paper.

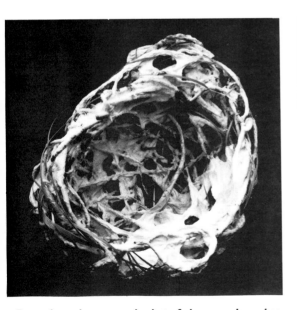

Frosted; random woven basket of elaeagnus branches
and handmade papers.

Nests of Willow; detail.

Plant Charts

Common Name: *Agave, sisal*
Botanical Name: *Agave sp.*
Where It's Found: *Arid regions of Southwestern U.S. and Mexico. Several species are houseplants.*
When To Harvest: *Fall*
How To Collect: *Cut leaves at the base of the plant.*
Part To Use: *Leaves*
Preparation: *Pound the leaves, using rubber gloves because the pulp may be an irritant; soak overnight; cook two to three hours with sodium carbonate.*
Miscellaneous: *Agave has been used since prehistoric times for cordage.*

- -

Common Name: *Blackberry, dewberry*
Botanical Name: *Rubus sp.*
Where It's Found: *Throughout the U.S.*
When To Harvest: *Spring for young shoots; fall for mature vines.*
How To Collect: *In the spring, prune new growth tips (roughly the top five inches); in the late fall, prune canes to desired length, leaving ample for next year's berries.*
Part To Use: *All of the tender new growth tips; inner bark from the canes.*
Preparation: *Cut canes into usable-sized pieces (a size that will fit into your papermaking cook pot); to remove bark, steam canes about 20 minutes; peel immediately, using a vegetable peeler or knife; now the inner bark can be cooked with sodium carbonate; the bark will take longer than the new shoots.*
Miscellaneous: *Blackberry shoots are dye plants, giving the paper a soft gray color.*

- -

Common Name: *Bulrush*
Botanical Name: *Scirpus sp.*
Where It's Found: *Marshes and ditches throughout the mainland U.S.*
When To Harvest: *Late summer after the seed head has formed.*
How To Collect: *Cut stalk at base of plant.*
Part To Use: *Stalks*
Preparation: *Cut stalks into short, usable pieces; crush stalks with meat tenderizer and soak four hours in clear water prior to cooking; cook two hours with sodium carbonate.*
Miscellaneous: *The seed heads can be added to the blender during the last 15 seconds of beating or added directly to the vat; the paper is a light tan color; bulrush is also a good basketry fiber.*

- -

Common Name: *Canna lily*
Botanical Name: *Canna indica*
Where It's Found: *Gardens throughout the U.S.*
When To Harvest: *Late summer or after a killing frost.*
How To Collect: *Cut leaves from stem about one inch above the bulb.*
Part To Use: *Leaves*
Preparation: *Scrape green leaves to remove outer skin; no need to scrape dry leaves; soak leaves (dry and green) in water four hours prior to cooking; cook three to four hours with sodium carbonate.*
Miscellaneous: *The paper is a light tan color.*

Common Name: *Cattail*

Botanical Name: *Typha latifolia*

Where It's Found: *Marshes and ditches throughout the mainland U.S.*

When To Harvest: *Late summer after the seed head has formed.*

How To Collect: *Cut leaves at base.*

Part To Use: *Leaves and stems*

Preparation: *Cut into short pieces; pre-soak in clear water for 24 hours; cook two hours with sodium carbonate.*

Miscellaneous: *Pollen from the seed can be added to the blender during the last 15 seconds of beating or floated in the van for added texture; the paper is a light tan.*

Common Name: *Cedar, juniper (Several species of juniper are also called cedar.)*

Botanical Name: *Juniperus*

Where It's Found: *Northwest, Oklahoma, Texas, Utah, New Mexico, and Mexico.*

When To Harvest: *Anytime from felled trees.*

How To Collect: *Peel off both the inner and outer bark from newly felled trees; this is easier in the spring when the sap is up; loose outer bark may be pulled from a living tree.*

Part To Use: *Bark from branches or trunk of tree; inner bark is the most desirable.*

Preparation: *This is a tough material; soak two to four days; if bark does not strip easily, steam the branches; bark can be boiled, simmered, and steeped at intervals over a two-day period, with total cooking time about eight to ten hours; periodically, more water and sodium carbonate should be added.*

Miscellaneous: *This material doesn't rinse clear; paper is a warm reddish-brown.*

Common Name: *Common reed*

Botanical Name: *Phragmites communis*

Where It's Found: *Marshy ground and ditches throughout the U.S.*

When To Harvest: *Late summer after the seed head is formed.*

How To Collect: *Cut stalks at the base of the plant.*

Part To Use: *Leaves, stalks, and plumes.*

Preparation: *Cut stalks into usable size and soak four hours; cook two to three hours with sodium carbonate.*

Miscellaneous: *For dye, the plant material is brought to a boil and left to steep overnight; cooking time is about 20 minutes; strain and save dye liquid; the paper is a golden tan; the plumes can be floated in the vat for an interesting texture.*

Common Name: *Corn husks*

Botanical Name: *Zea mays*

Where It's Found: *Farms and gardens throughout the U.S.*

When To Harvest: *Summer, after the crop has been harvested; in the Southwest, cornhusks can be purchased in grocery stores for tamales.*

How To Collect: *Remove husks from corn ear; cut stalk at base of plant.*

Part To Use: *Leaves, husk, stalk*

Preparation: *Cut into usable pieces, separate husks, soak in clear water 24 hours prior to cooking; cook two hours with sodium carbonate.*

Miscellaneous: *Paper made from dried husks, stalks, and leaves is a greenish-cream color; some Indian corn has wonderfully colorful husks; paper made from the purplish husks is a mauve color.*

Common Name: *Curly dock, broad-leaf dock*
Botanical Name: *Rumex crispus*
Where It's Found: *Along the roadside in temperate zones in the U.S.*
When To Harvest: *After the seed head has formed.*
How To Collect: *Cut the stems at the base of the plant.*
Part To Use: *Leaves, stems, and seed heads*
Preparation: *Cut into usable-sized pieces and soak overnight; cook two hours with sodium carbonate.*
Miscellaneous: *For dye, simmer for 30 minutes, strain and save dye liquid; return the plant material to cook for an hour and a half with sodium carbonate; the color of the paper is golden tan; add abaca for added strength.*

Common Name: *Daffodil*
Botanical Name: *Narcissus sp.*
Where It's Found: *Gardens throughout the U.S.*
When To Harvest: *After the foliage dies down.*
How To Collect: *Cut leaves at base of plant.*
Part To Use: *Leaves*
Preparation: *Soak 24 hours prior to cooking; cook an hour and a half with sodium carbonate.*
Miscellaneous: *Add abaca for stronger paper; the paper is a light tan color.*

Common Name: *Day lily*
Botanical Name: *Hemerocallis fulva*
Where It's Found: *Flower gardens throughout the U.S.*
When To Harvest: *Leaves may be harvested any time during the growing season.*
How To Collect: *Gather the dry leaves at the base of the plant.*
Part To Use: *Leaves for the paper; flowers can be added for color and texture.*
Preparation: *Cut leaves to a usable length; soak 24 hours; add sodium carbonate, bring to a boil, then simmer approximately two hours.*
Miscellaneous: *Flowers are best gathered towards the end of the day; they can be used for dye; cook separately without the sodium carbonate; add liquid to the vat for color; flowers can be chopped and added to the vat for color and texture.*

Common Name: *Eucalyptus*
Botanical Name: *Eucalyptus sp.*
Where It's Found: *California*
When To Harvest: *Anytime*
How To Collect: *Gather bark that has fallen from the tree.*
Part To Use: *Bark*
Preparation: *Soak bark in clear water 24 hours prior to cooking; cook two to three hours with sodium carbonate.*
Miscellaneous: *The paper is a reddish brown color.*

Common Name: *Fig*
Botanical Name: *Ficus*
Where It's Found: *Hawaii, Gulf states, Southeastern states, California, Arizona, New Mexico, and Texas.*
When To Harvest: *Summer or fall.*

How To Collect: *Prune according to needs of the tree.*

Part To Use: *Inner bark from the branches*

Preparation: *Remove leaves; strip fiber from the woody core immediately after cutting branches; if branches are dry, steam to loosen the bark; cook three to four hours with sodium carbonate; after cooking, pull fiber apart and cut into 1/2-inch pieces before blending.*

Miscellaneous: *Paper is a tan color.*

Common Name: *Gladiolus*

Botanical Name: *Gladiolus sp.*

Where It's Found: *Gardens*

When To Harvest: *After flowering, when the foliage dies.*

How To Collect: *Cut the foliage off 1/2 inch above the bulb.*

Part To Use: *Leaves and stalks*

Preparation: *Cut into usable pieces, soak 24 hours prior to cooking; cook two hours with sodium carbonate.*

Miscellaneous: *Paper is a light tan color.*

Common Name: *Goldenrod*

Botanical Name: *Solidage sp.*

Where It's Found: *Dry open fields throughout the U.S.*

When To Harvest: *Late summer or fall*

How To Collect: *Cut stems at base of plant.*

Part To Use: *Flowers, leaves, and stems*

Preparation: *Cut into usable-sized pieces and soak overnight; cook for one to two hours with sodium carbonate.*

Miscellaneous: *For dye, cook 45 minutes, strain, and save dye liquid; return plant material to cooking pot with sodium carbonate to simmer for one hour; the color of the paper is a pale greenish-beige; add abaca pulp for strength.*

Common Name: *Iris*

Botanical Name: *Iris sp.*

Where It's Found: *Gardens throughout the U.S.*

When To Harvest: *Fall or after frost.*

How To Collect: *Cut dry leaves at the base of the plant; when the bulbs are divided, green leaf tips can be cut.*

Part To Use: *Leaves*

Preparation: *Scrape fresh leaves; no need to scrape dry leaves; soak in clear water 24 hours prior to cooking; cook two to three hours with sodium carbonate.*

Miscellaneous: *Paper is a light tan color.*

Common Name: *Kudzu*

Botanical Name: *Pueraria lobata*

Where It's Found: *Southern states*

When To Harvest: *Summer*

How To Collect: *Cut vines in usable lengths.*

Part To Use: *Vines*

Preparation: *Cut pencil-thick vines eight to ten feet long; coil them and soak 24 hours; cook three hours with sodium carbonate; cut into short pieces before putting into the blender.*

Miscellaneous: *The paper is a light green with a translucent quality; it is illegal to transport kudzu across some state lines.*

Common Name: *Milkweed*

Botanical Name: *Asclepias sp.*

Where It's Found: *Roadsides and fields with dry soil throughout the U.S.*

When To Harvest: *After the seed pods are formed.*

How To Collect: *Cut stems and strip for bast fiber.*

Part To Use: *Internal (bast) fiber from the stems*

Preparation: *Cut stems into usable pieces; soak 24 hours prior to cooking; cook stem fibers with sodium carbonate for an hour and a half.*

Miscellaneous: *Blossoms, leaves, and stems may all be used to obtain a very light yellow dye; the fluffy seeds may be added to the vat for textural interest.*

Common Name: *Okra*

Botanical Name: *Hibiscus esculentus*

Where It's Found: *Gardens; frozen okra pods can be obtained from the grocery store.*

When To Harvest: *Fall*

How To Collect: *Pull the plants up, roots and all; plants may be left to ret in the garden after the plant dies; the inner fibers of the stem will break down, making the usable fibers easier to obtain.*

Part To Use: *All parts can be used, including the pods.*

Preparation: *Cut into manageable-sized pieces; if dry, crush the pods, stems, and roots; steam and strip fibers from the stem; cook two hours with sodium carbonate.*

Miscellaneous: *Okra is in the same plant family as the Hibiscus manikat; the root of this plant is the part used for a formation aid in Japanese papermaking; I have added a cup of frozen okra to other fiber pulps in the cooking stage as a formation aid.*

Common Name: *Osage Orange*

Botanical Name: *Maclura pomifera*

Where It's Found: *South Central U.S.*

When To Harvest: *Anytime the tree is pruned, felled, or uprooted.*

How To Collect: *Cut roots and strip bark.*

Part To Use: *Roots, tree bark, external, and internal*

Preparation: *Cut branches and roots into a usable size; strip the bark; both outer and inner bark may be used; first cook one hour for dye, strain and save dye liquid; cook the roots and bark two hours with sodium carbonate; may need to cook longer or require a second cooking with fresh water and sodium carbonate.*

Miscellaneous: *Excellent dye in the tan to yellow-gold range.*

Common Name: *Pineapple*

Botanical Name: *Ananas comosus*

Where It's Found: *Hawaii*

When To Harvest: *Anytime; purchase at the grocery store.*

How To Collect: *Cut off the leafy top of the pineapple.*

Part To Use: *Leaves*

Preparation: *Scrape fresh leaves to remove the outer skin; soak in clear water for 24 hours prior to cooking; cook two hours with sodium carbonate.*

Miscellaneous: *The paper is a greenish-cream color.*

Common Name: Privet
Botanical Name: Ligustrum vulgare
Where It's Found: Gardens and hedges
When To Harvest: Anytime the hedge is trimmed.
How To Collect: Prune the branches; strip the leaves from the branches.
Part To Use: Leaves and the inner bark from the branches
Preparation: Steam the branches to remove the inner bark; cook hours with sodium carbonate.
Miscellaneous: Fresh leaves will produce a dull to bright yellow color; the leaves can be cooked separately to obtain the color; strain and save the colored liquid; add leaves to the bast fiber to be cooked for paper.

Common Name: Sagebrush
Botanical Name: Artemisia tridentata
Where It's Found: Western states
When To Harvest: Anytime the shrub is trimmed, but late summer is best.
How To Collect: Cut twigs at trunk or branch.
Part To Use: Leaves, twigs
Preparation: Cut to usable size; steam and crush twigs; soak 24 hours prior to cooking; cook two hours with sodium carbonate.
Miscellaneous: Sagebrush is a good dye plant; the dye color is a greenish-yellow; cook without the sodium carbonate to obtain the dye liquid; pour off and save dye liquid; return plant material to cooking pot with sodium carbonate and water to cook for another hour; add dye liquid to the vat.

Common Name: Sansevieria
Botanical Name: Sansevieria trifasciata
Where It's Found: Southern Florida; houseplant.
When To Harvest: Anytime
How To Collect: Cut leaf at base.
Part To Use: Leaves
Preparation: Scrape fresh leaves to remove outer skin; pre-soak in water for four hours prior to cooking; cook two hours with sodium carbonate.
Miscellaneous: Paper is a greenish-cream color.

Common Name: Sedge
Botanical Name: Carex sp.
Where It's Found: Ditches and marshes throughout the U.S.
When To Harvest: Fresh leaves and stems in the late summer.
How To Collect: Cut stems at base of plant.
Part To Use: Leaves, stems, and seed heads
Preparation: Soak 24 hours prior to cooking; cook an hour and a half with sodium carbonate.
Miscellaneous: The paper is a yellow-beige color; both the roots and leaves are used as a basketry fiber.

Common Name: Sumac, flame leaf, staghorn sumac.
Botanical Name: Rhus sp.
Where It's Found: Texas, Louisiana, eastward to Florida, northward to Minnesota, New York, and Rhode Island.

When To Harvest: *Fall*

How To Collect: *Pull leaf stems from the branches and trunk.*

Part To Use: *Leaves and stems*

Preparation: *Remove leaves from stems and cut stems into usable pieces; soak leaves and stems 24 hours prior to cooking; cook two hours with sodium carbonate.*

Miscellaneous: *Leaves gathered in the fall are rich colors to use in the layering technique or chopped and added to the vat; care should be taken that one does not experiment with any of the poisonous variety; poison ivy and poison oak are close kin to poison sumac; poison sumac has small gray berries; non-poisonous varieties have deep red berries; poison sumac likes a more moist habitat.*

Common Name: *Umbrella plant, papyrus*

Botanical Name: *Cyperus papyrus*

Where It's Found: *Gardens*

When To Harvest: *After the seed head has formed*

How To Collect: *Cut stem at base of plant.*

Part To Use: *Stem and leaves*

Preparation: *Cut into usable pieces and soak 24 hours prior to cooking; cook two hours with sodium carbonate; crush the stems; pull the fibers apart and cut into half-inch pieces before blending.*

Miscellaneous: *Papyrus as a writing surface was first made by Egyptians during the third century B.C.; the technique to make that type of paper is quite different; strips of the stems are placed side by side on cotton cloth and pressed to form a sheet.*

Common Name: *Wheat*

Botanical Name: *Triticum acstivum*

Where It's Found: *Farms throughout the Midwestern and Southwestern states.*

When To Harvest: *After the crop has been harvested.*

How To Collect: *Cut the stems at the plant's base.*

Part To Use: *Stalks, seed heads, and leaves*

Preparation: *Cut into usable pieces; soak four hours prior to cooking; bring to a boil and simmer for an hour and a half with sodium carbonate.*

Miscellaneous: *Mix abaca pulp with the wheat pulp for added strength; the paper is cream colored with bits of gold.*

Common Name: *Willow*

Botanical Name: *Salix sp.*

Where It's Found: *Different species throughout the U.S.*

When To Harvest: *Spring or summer*

How To Collect: *Cut branch off at the trunk or next branch.*

Part To Use: *Bark and leaves; I use both the inner and outer bark.*

Preparation: *The bark can be easily stripped from the freshly cut branches with a knife or vegetable scraper; after the branch dries, cut into manageable pieces; steam 20 to 30 minutes, then strip the bark immediately; soak 24 hours; cook two hours with sodium carbonate.*

Miscellaneous: *Pressed leaves and fluffy seeds are excellent objects to sandwich in between two thin sheets in the layering technique; leaves and seeds may be added to the vat.*

Common Name: *Yarrow*

Botanical Name: *Achillea millefolium*

Where It's Found: *Roadsides, fields, gardens; widely distributed across most temperate zones in the U.S.*

When To Harvest: *After it blooms.*

How To Collect: *Cut stems at the base of the plant.*

Part To Use: *Stems and leaves*

Preparation: *For dye, cook one hour; strain, and save dye liquid; cook plant material one hour longer with sodium carbonate.*

Miscellaneous: *Wonderful garden plant; it can be used in fresh or dried arrangements, giving off a pleasing, pungent fragrance; the flowers and fern-like leaves may be added to the vat; the dye color is a golden beige.*

- -

Common Name: *Yellow sweet clover*

Botanical Name: *Melilotus officinalis*

Where It's Found: *Widely distributed throughout North America.*

When To Harvest: *When the plant blooms.*

How To Collect: *Cut at base of the plant.*

Part To Use: *Stems, leaves, and flowers*

Preparation: *Cut into usable-sized pieces; for dye, cook covered with water one hour; let plant fibers cool before straining; save the dye liquid; cook fibers about one hour with water and sodium carbonate for pulp.*

Miscellaneous: *The dye color is a light yellow.*

- -

Common Name: *Yucca*

Botanical Name: *Yucca sp.*

Where It's Found: *Throughout the U.S.*

When To Harvest: *After the seed pod has formed.*

How To Collect: *Cut leaves at the base of plant.*

Part To Use: *Leaves*

Preparation: *Scrape fresh leaves to remove outer skin; soak in clear water for 24 hours prior to cooking; cook two hours with sodium carbonate.*

Miscellaneous: *The paper is a light greenish-cream color.*

- -

Pine Paper Pot, Cloud Basket Series; stitched and coiled basket of cotton linters with pine needles, waxed linen, pine needles, dye; 7" x 11" h; photo: Judy Mulford.

JUDY MULFORD

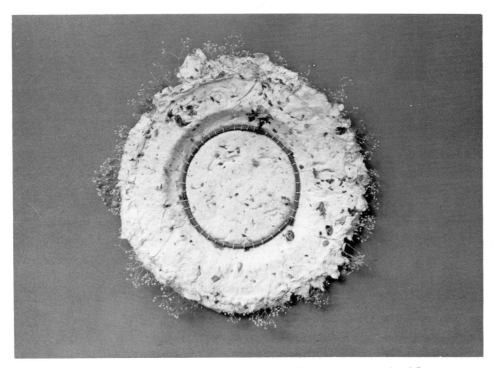

Potpourri Paper Plate, Cloud Basket Series; paper linters, potpourri, dried flowers,
pine needles, and waxed linen; 14" x 1-1/2" h.

Cloud or Spirit Baskets

Making containers with handmade paper gives me a joyous, spiritual high. The baskets are light and airy, fragrant and colorful, and almost instant. I love them because they are the complete opposite of my "soul" baskets, which are tedious, dark, and heavy.

There is a duality in my work and my life. The paper "cloud" baskets represent the color and joy and spontaneity that I wish I had more of in life. They are the "butterfly" baskets of my smile.

The pine needle/clay "soul" baskets represent my inner feelings of frustration and fear and my sensitivity. They are somber and repetitious. They are the "bee" of my being.

The materials I use are white paper linters, combined with dye, paint, wax, potpourri, ribbons and thread, and recycled brown grocery bags. I either layer the pulp over a basket, or press layers of sheets into my dishes and pans. I dry them in the oven or the microwave.

Untitled; random weave with grapevine, brown paper bag pulp, and eucalyptus bark; 8" x 7" h.

Cloud Basket with Pine Needles; dyed pine needles and cotton linters with boiled pine needles; 4" w x 13" h.

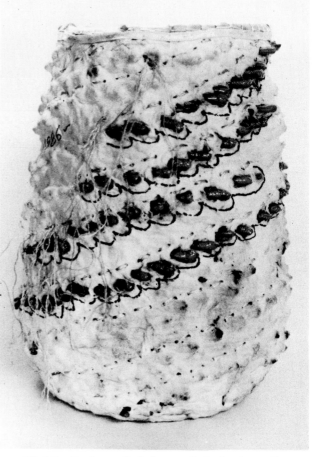

Wedding Basket; coiled pine needle basket layered with dyed cotton linter pulp; 5" x 6" h.

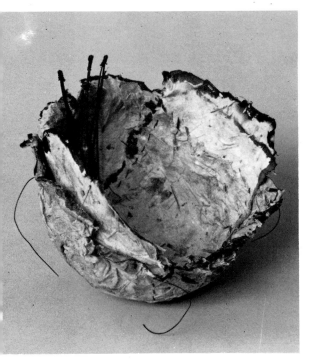

Cloud Basket Series; brown paper bag pulp, eucalyptus bark, pine needles, dye, and stitching; 8" x 7" h.

Random Weave; willow dipped in brown paper bag pulp and layered with sheets; 6" x 2-3/4" h.

Mini Random Weave and Paper; mysterious vine dipped in brown paper bag and cotton linter pulp; 2-1/4" x 1-1/2" h.

Untitled; brown paper bag sheets with waxed
linen in between, dipped in brown
hot wax; 4" x 1-3/4" h.

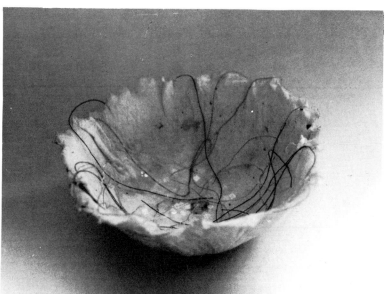

A true Cloud basket; two layers of dyed cotton linter paper with
thread and glitter in between; 5-1/2" x 2" h.

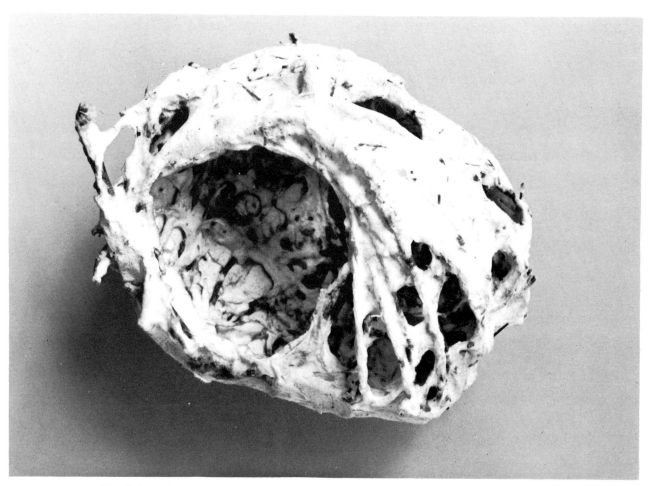

Untitled; random weave in willow covered with pine needles and cotton linter paper; 9" x 5-1/2" h.

Pine Needle Cloud Basket; pine needles, cotton linters, wax, dye, and waxed linen; 6" x 6" h.

Random Weave Wasp's Nest; mysterious vine with brown paper bag pulp layered on;
7" x 5" h.

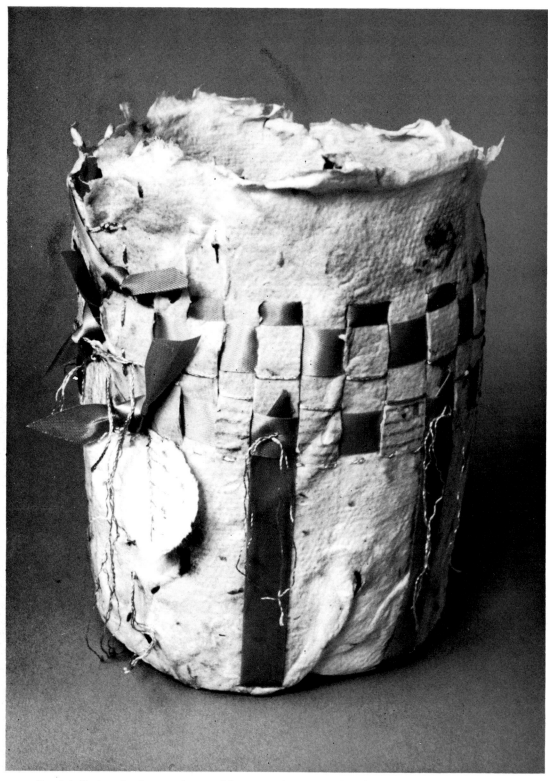

Boudoir Basket; dyed cotton linters with potpourri, woven with ribbon, stitched with embroidery floss, and embellished with markers; 6" x 8" h.

Stacked Cylinder; 6" x 6" x 12"; photo © 1987 North-South Photography.

ALICE WAND

Bowl; raffia and round reed trim; 13"d.

I began working with paper pulp about 10 years ago when I took a workshop with Elaine Koretsky of Carriage House Paperworks in Boston. She is very interested in alternative fibers for papermaking, such as Manila hemp (abaca) pulp from the banana plant family. I like this pulp as well, and have used it almost exclusively over the years. It is very tough and makes lovely surface textures.

I soon tired of making sheets of paper, and because of my fine arts background, I began experimenting with various odd shapes and three-dimensional possibilities. Out of this experimentation came decorative fans, boxes, weed pouches, window circles, and eventually bowls, baskets, and saucers.

The bowls and baskets are formed over beach balls; the saucers are formed over children's round plastic sleds. All the pulp contains sizing and is colored with organic pigments to make it as lightfast as possible. I use round reed and hand-dyed raffia in many of the pieces.

Recently, my work has become more elaborate and ornate, with less emphasis on production work. I've begun to use embedded cut strips of layered pulp—which I call "squiggles"—in the surface of the object. I'm having a great deal of fun with this technique, and I imagine I will continue experimenting with this idea for a while.

Handmade paper box with lid; Manila hemp pulp, raffia, and round reed trim; 7" x 7" x 12".

Box; handmade paper, Manila hemp pulp, sprayed with paint; 6" x 6".

Handmade paper bowl with squiggles; raffia trim; 15"d; photo© 1987 North-South Photography.

Handmade paper basket with handle; Manila hemp pulp, round reed, and raffia trim; 15" d.

Handmade paper box with lid; Manila hemp pulp with raffia and round reed; 6" x 6" x 12".

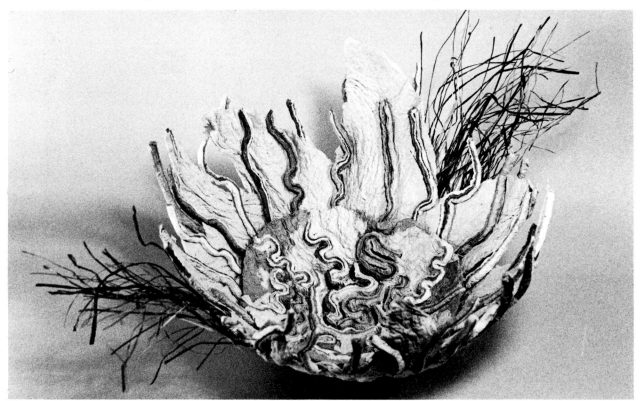

Open bowl; handmade paper; 18"d.

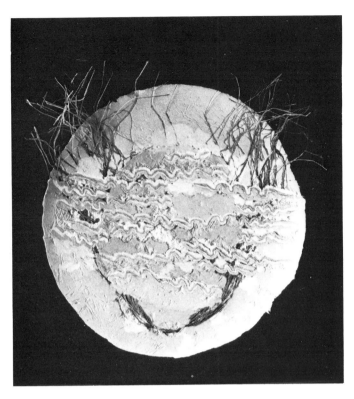

Handmade paper wall saucer; 21"d; photo© 1987 North-South Photography.

Handmade paper box with lid; raffia and round reed trim; 7" x 7" x 10"; photo© 1987 North-South Photography.

Nest of Mini Dodo Bird; cast handmade paper, Procion dye, raffia, porcelain eggs, acrylic pigment, and river pebbles.

PAM BARTON

Nest for Butterflies, Nest Tropicana, and Nest of Mini Dodo Bird.

One of the countless pleasures of working with fibers is that I can go from the very sublime to the absolutely ridiculous—all with a clear conscience. By no means do I limit myself to just natural fibers, as most elements can play a role in my ceremonial process at one time or another. Just being alive energizes me, and I am inspired by my surroundings: the sun, the moon, the rain, trees, plants, rocks, and soil, the glitter and bulk in industry, my husband…and by being me.

It may seem ambiguous to cast a basket and weave it. But this is precisely what this project entails. Part of the basket is cast paper, and part of it is woven. Yet the two components are literally interwoven to form one basket. With some of the weaving elements concealed in the cast portion, a bit of mystery is created for the viewer.

The process is quite involved and very time-consuming, but for me, combines most of my loves: clay (substituted by cast pulp), vessel forms, basketry, fiber, paper, and textiles. The slow process of construction allows time to fantasize about each piece so that the finished creation becomes a very personal extension of myself.

Before embarking on this venture, review the instructions thoroughly, and check the supply lists which are at the beginning of each section.

Preparing the Mold

Supplies:

- lots of enthusiasm, love for baskets, a happy heart, and willing hands
- a mold on which to cast
- foil
- plastic wrap
- masking tape

1. Select a form suitable for the lower portion of your basket. It must be of a material that will retain its shape in moisture and heat, such as a bowl made of ceramic, wood, plastic, metal, or glass. Paper pulp will be cast on this form.

2. The base of the form must be smaller, or of the same size, as the top so that the cast piece can be removed easily when dry. If the form has a hollow foot, I usually fill the foot with clay to create a solid base. Next, I cover the bottom of the foot with card board of the same shape, and then cover the entire form with a layer of foil.

3. Now cover the foil with a layer of plastic wrap. The foil protects the form from the binding agent, while the plastic makes it easier to unmold the dried casting. Use masking tape to keep the plastic wrap taut over the form, but don't tape on the surface area where pulp will be cast because dried casting sticks to tape. I use a piece of masking tape around the circumference of my mold as a guide for the top edge of my basket (the bottom edge while casting). With the form covered, you've prepared your mold and may proceed to the next step.

Making the Base

Any fiber that will not become deformed by heat or water during the casting and drying stages can be used. Good examples are seagrass, raffia, cotton, or nylon cordage. Round reed makes a good core. I use seagrass for my coil (I like its flexibility), and usually raffia for my wrap, warp, and weaver elements. Both hold up well in moisture and heat and are readily available.

Supplies:

- fibers of your choice
- craft needle with a large eye

1. With core fiber, coil two, three, or more rings in a size that best suits the shape of your mold.

2. Do a figure-eight stitch with a weaver to connect the rings. I like to add an offbeat twist to my base.

3. Then add warp elements. The length of the warp elements should be twice the height of the basket, plus four inches for finishing. Each length becomes two warp elements. Thread these individually through the top row of the coiled base with your needle. Use as many warp elements as desired to enhance your basket, but not so many as to fill the spaces between warps, which allow the two layers of pulp to bond.

4. With a weaver, do four or more rows of twining, shaping the base to fit the bottom of your basket. Now move on to casting.

Preparing the Base

Cover the selected form with foil and plastic wrap. Use masking tape, if necessary, to keep the wrap taut over the form, but avoid laying tape on any area over which pulp will be cast.

Coil three or four rows of seagrass. Using a figure-eight stitch, weave the coiled elements together to form a circle.

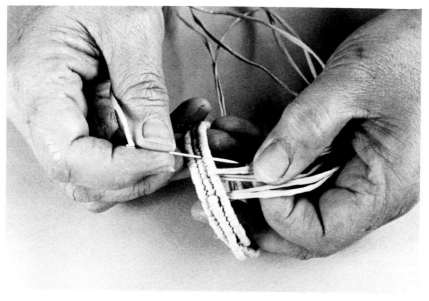

Add warp elements, spaced evenly apart.

Preparing the Pulp

Rather than purchased linters, I usually use fiber scraps from my basketry and papermaking to make my pulp. This involves a lot of time for cooking and beating the fibers, but allows me to use my leftovers. The process can be simplified by using linters (commercially prepared fibers in sheet form or loose), or by recycling various types of paper or paper stock, and then adding embellishments and textures. I've included both recipes, and each recipe yields two 3" x 5" basket bottoms.

Method I

One cotton linter measures 24" x 36", and yields 18–19 cups of pulp. Each 3" x 5" basket bottom requires one cup of pulp.

Supplies:

- Paper stock—old prints, watercolor or printmaking paper, newsprint, brown paper bags, scraps of Japanese or other handmade paper
- cotton linter (I recommend second cut cotton linters which cost less and are easy to work with)
- texture embellishments (dried leaves, ferns, grasses, threads, found objects, etc.)
- a blender
- sharp scissors
- measuring cup
- colander or strainer
- fine screen mesh (for use in sink drain)
- fine mesh fabric (large enough to line the colander or strainer; the nylon or polyester used for silk screening is great)
- a gallon-size plastic bowl (for mixing pulp)
- a binding agent (matte or gloss acrylic polymer medium, wheat paste, or vinyl wallpaper paste; my favorite binding agent is Polyadam 1. See Suppliers list.)

1. Cut or tear the selected stock into 1/2- to 3/4-inch pieces until you have two packed cups of linters, or four cups of other paper stock, or a mixture of the two.

2. Add up to one-half cup of texture embellishments which have been cut into 1/2-inch pieces to the torn stock.

3. Cover the mixture with water and allow it to soak for at least one hour, but no longer than overnight. (If you allow it to soak too long before draining, it may sour.)

4. Drain the water.

5. Place the wet mixture, 1/4-cup at a time, into the blender and fill with cold water to within two inches of the top. DO NOT OVERFILL. Blend at the lowest speed for 15 seconds (I count to 15), then at the highest speed for five seconds, then back to the lowest speed for five more seconds. I've been told by a small appliance repairman that this is easier on the blender than to switch it off from the highest speed. If the blender runs hard and rough, divide the pulp mixture in half, add water to within two inches of the top, and blend each half separately. Proceed accordingly until all of the wet mixture has been blended.

6. Pour the blended pulp into a colander or strainer lined with the fine mesh fabric. The fabric liner is essential to prevent clogging the sink drain.

7. Allow the pulp to drain for about one hour. You can help the draining process by shaking the lined strainer back and forth.

8. Dyeing or coloring the pulp can be done now. See the instructions for how to color the pulp on page 58.

9. If using immediately, place the dyed pulp in a plastic container. Add 1/3-cup binding agent and mix well. Then return the pulp to the lined strainer.

10. If not using immediately, the pulp can be stored in the refrigerator up to two weeks. To store, put pulp in a plastic container, then press a piece of plastic wrap directly over the top of the pulp. Cover the container with a tight-fitting lid and place in refrigerator. When ready to use, allow the pulp to reach room temperature, and then add 1/3-cup binding agent, mix well, and place in the fabric-lined strainer.

11. Wash and dry blender and all other materials. You are now ready to start casting the bottom of your basket.

HELPFUL NOTE: If working in the kitchen or at a drain with plumbing, insert a piece of fine screen mesh in the drain to catch the loose pulp and prevent clogging.

Nest Tropicana; cast handmade paper, Procion dye, raffia, embroidery thread, and beads.

Method II (using natural fibers and scraps)

This is my preference because I always have an abundance of fibers and scraps.

Supplies (these are in addition to those listed for Method I):

- stainless steel pot
- a mallet (for beating the fibers)
- a heavy chopping board (to beat the fibers on)
- household washing soda or lye
- bamboo, wooden tongs, or chopsticks
- a long-handled wooden spoon
- three pounds (dry weight) fibers (I use the inner bark of paper mulberry, hibiscus and wickstroemia mixed or singularly, or flax; and add whatever happens to be around at the time, such as raffia, jute, ti leaves, cattail stems and leaves, dried fern, or sisal.)

1. Cut the fibers into one-inch lengths. Cover with water and soak for 24 hours.

2. Place the wet fibers in a stainless steel pot and add cold water, by the quart, to cover the fibers. For each quart of water, add one tablespoon washing soda or lye. **Follow all precautionary measures that apply when using chemicals.** Bring mixture to a slow boil, reduce heat to low, cover, and simmer at least three hours. Stir occasionally with a wooden spoon or stick. Cook in a well-ventilated area. Fibers are ready if they separate easily when squeezed between thumb and finger. (Please use gloves!) Remove fibers with tongs or chopsticks to avoid skin contact with lye mixture, and rinse in cold water before testing.

3. Allow the fibers to cool and then rinse thoroughly in cold water. It will not hurt the fibers to sit overnight in the cooking solution.

4. The fibers are now ready to be beaten. Beat the fibers until they are no longer fibrous. By hand, it will take about two hours. With a Hollander, it will take 15 to 20 minutes. I use my tapa beaters on my tapa-making anvil; both are made of wood. A wooden mallet or wooden meat tenderizer, or even a small baseball bat and a chopping board will serve the purpose. Beating fibrillates the fibers and roughens their surface. A rough surface takes up water more readily and helps achieve better bonding.

5. Soak the beaten fibers for 24 hours in cold water so they will be more malleable in the blender. To the soaking fibers, I add up to three cups of handmade paper (Japanese or my own) torn into one-inch pieces, and up to one cup of texture embellishments.

The rest of the process is identical to that described in Method I, starting at step 5.

Coloring the Pulp

I have used pearlescent and aqueous pigments and fiber reactive dyes with satisfactory results.

Supplies:

- a plastic container large enough to hold pulp with room to spare
- wooden spoon
- pigments of your choice

  - *Pearlescent pigments add marvelous sparkle and excitement.*
  - *Aqueous dispersed pigments produce intense color and require a retention aid. (Both pearlescent and aqueous pigments are safe and very easy to use.)*
  - *Dyes take more time than pigments and produce more subtle colors.*

Punk Roc Nest; cast handmade paper, Procion dye, wool, embroidery floss, metallic thread, and river pebbles.

   — *Acrylic pigments (liquid and dry) may also be used, but with discretion. Colors are more intense, but artificial, and they must be thinned, smoothed in warm water, and strained before mixing with pulp. If the pigment is too thick, it will coat the fibers and prevent a good bond.*

   — *Read and carefully follow written instructions on the package.*

Casting the Bottom

Supplies:

- a base on which to support your mold during casting
- sponge
- prepared pulp
- additional warp elements (if necessary)

1. Place your mold, bottom side up, on the base.

2. Working with small portions, squeeze some liquid out of the condensed pulp with your fingers and press onto the mold until it's covered with an even thickness of pulp. Sponge frequently to absorb excess liquid during the casting process. If the layer of pulp is too thick, it will take too long to dry and may develop mold.

3. Dampen the warp elements of the base. Then position the base in place on the wet casting and space the warp elements evenly, allowing casting to show between them to ensure proper bonding of the two pulp layers. If it appears that you will need additional warp elements, and if space permits between existing warps, they can be added when you have covered the bottom third of the mold. Lay the dampened additions in place on the first layer of cast pulp and cover them with the second layer, being certain that at least an inch and a half of the warp is covered with the wet pulp.

4. Apply a second pulp layer of even thickness over the entire mold, including fibers. When applying the second layer, exert moderate pressure to ensure bonding between the layers. Sponge frequently. After applying the second layer, give it a good sponging and set it to dry. DO NOT allow the first cast layer to dry before applying the second. Pulp should be of the same moisture content to ensure proper bonding. (Wet pulp applied to dry pulp tends to separate during the drying stage.)

5. Clean up your space and equipment.

HELPFUL NOTES: Have a bucket of water handy for rinsing your pulpy hands and equipment, and a place other than down the drain to empty it.

Drying and Finishing the Bottom

Supplies:

- hand-held hair dryer
- bee's wax
- container and means for melting wax
- inexpensive brush for applying wax
- coloring pigment (if desired)

1. Allow the cast piece to dry thoroughly on the mold. Keep the warp elements hanging as straight as possible. I live in a rain forest area, so I dry my pieces in front of the fireplace or heater, rotating them frequently. I look forward to a microwave oven which should expedite drying tremendously. (My husband fears that in getting one, he'll only see baskets and paper coming out of it!) Check the manual for your microwave oven to see if you can use aluminum foil in it.

2. Remove the cast piece from the mold when the outside is dry and hard to the touch. If the inside is still damp, allow it to dry thoroughly. You can help it dry with the hair dryer, working from the center to the edges.

3. If the cast piece does not feel firm or stiff enough (it should have the firmness of a strong cardboard box), the following can be done: Melt bee's wax and brush a thin coat on the inside of the cast piece. With a hot hair dryer, melt the wax, which will then absorb into the paper piece. Let it cool and harden; then repeat the process on the outside. The wax will slightly darken the color. If a darker color is desired, a coat of water-based pigment can be painted on before applying the wax. Allow it to dry thoroughly; then proceed with the application of the bee's wax as directed above.

The dried cast piece should be about 1/8-inch thick—preferably less. In the sun it will take at least two days to thoroughly dry. With your basket bottom dried and firm, you are ready to weave the top.

Nest for Butterflies; cast handmade paper, Procion dye, raffia, cotton yarn, polyester, and nylon.

Making the Base

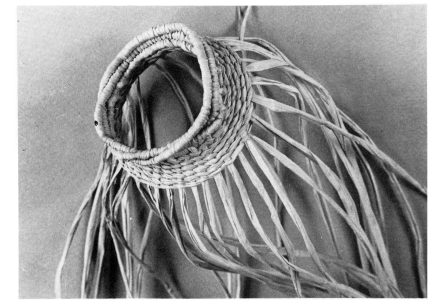

Do three or four rows of twining. I like to add an off-beat twist to my base.

Squeeze the liquid out of the condensed pulp with your fingers, and press onto the form to cover it with an even layer of pulp.

A layer of pulp of even thickness.

Dampen the warp elements. Place the base on the wet cast form, and evenly space the warp elements. Leave sufficient space between the elements to ensure proper bonding of the two layers of pulp.

Apply another even layer of pulp to cover the entire form, including the warp elements. Follow the process for the first layer, exerting moderate pressure to ensure bonding between layers.

Weaving the Top

Supplies:

- renewed enthusiasm (you are on the last segment of the project)

- fibers and/or yarns for weaving upper portion of basket

Use any weaving technique that will accomplish the shape you fancy. I'm partial to twining these basket forms, combining a variety of fibers with beads, feathers, and whatever else happens to be in my studio. It is during this segment that we really sing, laugh, and talk with each other—my basket and I. Now take your finished basket, place it in a well-lighted spot, and stand back and bask in the glow of your creation.

Remove the cast piece from the form. You are now ready to weave the remainder of your basket.

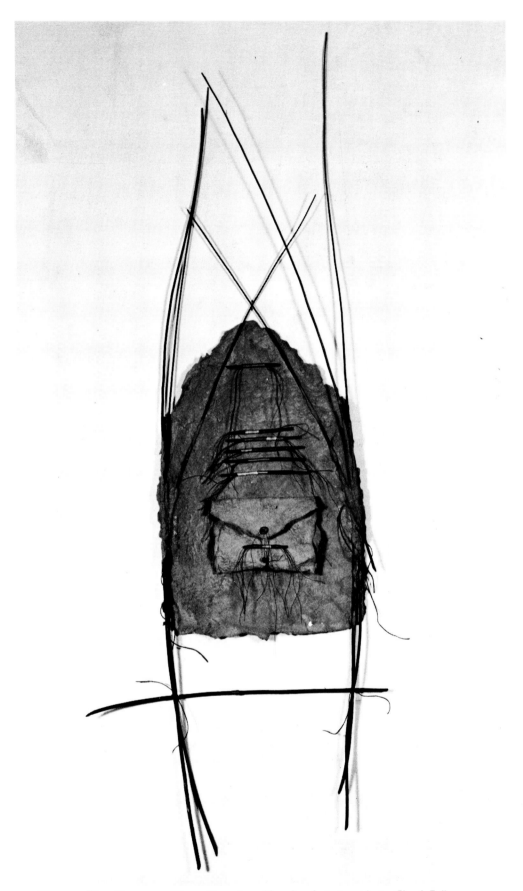

Message #2; willow, waxed linen, thread, and handmade paper; photo: Chuck Fulkerson.

MARY LEE FULKERSON

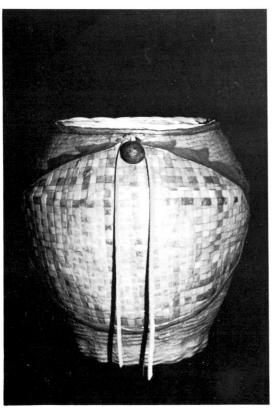

Moon Bless; 10" x 10" x 12"; photo: Carol Bakken.

Ghost Dance; 12" x 14" x 16"; photo: Carol Bakken.

Seeking Spirituality

Many artists are archaeologists, digging for their buried symbols. I know I am, and in a recent series of baskets I've discovered some of those symbols. Combining willow and reed with handmade paper, I try to refer to an earlier, spiritual time. I hope there is something here for some of you. After all, many footsteps take the same path, but they don't necessarily end up in the same place.

I live near the Pyramid Lake Indian Reservation, an area inhabited for 4,000 years by tribal people. The lake is still considered sacred by many. Near the pyramid, surveying the salt water which she created with her tears before turning to stone, is the Stone Mother and her Basket. She whispers of the spirituality of an earlier time. Today the earth is newly revered. It is Gaea, our Mother—our home and sustenance. For thousands of years it has been a part of religious and ritual thought and practice.

As an ancient craft using materials of the earth, many basket artists are involved in ritual as they weave, coil, or twine in studied repetition. The basket artist begins to create a rhythm. Contemplation sets in. Weaving with willow, reed, splint, or leaf, the invisible thread of thought and feeling flows through your fingers and into your basket. Are you aware of this when you work? If so, your ritual has begun. You are questing your symbols.

My technical approach involves four steps: visualization, shaping, adding handmade paper, and embellishing.

Fertility Jug

Visualization

For many Native Americans, images are a way of celebrating mystery, not explaining it. This "celebration theory" is my approach when visualizing a completed basket. *Song Keeper* reflects the belief that a song is made to be heard by those around it and then disappear. I visualized a shallow basket that opened to the sky and stretched outward.

Fertility Jug speaks of this legend: A woman wanted children but her husband did not. She put feathers in a water jug, and on the way home the feathers turned into babies. But the lid came off, and all the babies fell out except one. He was a Paiute and too fat to get out! My mind's eye saw the early water carriers, fat to hold a fat baby.

Ghost Dance illustrates an old Native American religious dance reflecting a never-ending hope for peace and wholeness in the world. And while *Dhurrie Basket* designs are taken straight from the rugs of the same name, there is a witches' coven on the steps conducting a moonlight ritual.

Your conceptualization would be completely different. But as long as there is a reason for it, there will be substance to your piece. Technique, shape, color—everything has to do with the source story.

Shape

In *Song Keeper,* flat reed was tied to another basket until dry. Part of *Fertility Jug* went over a football, and *Legend Jug* was formed around a wine bottle. They were tied with rag strips and dried to shape overnight before finishing.

The parts make up the spirit of the whole. Willow is mystically referred to by herbalists when calling up moon magic or wishing magic. When cutting willow, I bury a small offering of thanks nearby. I drop a piece of juniper in the soaking water for protection against negative vibrations.

Paper

Paper is ideal for reproducing an ancient look. For *Wano* to look weathered and tattered, I couched (transferred) the paper sheets directly onto the basket. The very wet

Song Keeper; reed, waxed linen, handmade paper, and acrylic paint.

and fragile pulp tore somewhat. Drying, it shrank in the most wonderful way, grasping at the basket skeleton. Mistakes are easily fixed, as the pulp can simply be scooped off and the couching process repeated.

I put pulp into a plastic container and mixed in acrylic paint. I smeared this over *Legend Jug*. I wanted the look of clumpy clay, and I was able to expose open spaces so it looked "found".

Fertility Jug is twined and then covered with papier-mâché to refer to ancient water jugs covered with pine pitch. Also, I wanted to add another layer to this piece in keeping with the layers of its legend.

Most of my recent work has involved liberal use of papier-mâché. There seems to be something quite wonderful about covering an ordinary basket with such ordinary material to produce extraordinary results. Bits of pulp and papier-mâché can be used to secure shells, feathers, etc., to the pieces. My method of making paper is very primitive, but it works well for my baskets.

Papier-Mâché

Instant papier-mâché is available at craft shops. To make your own, tear newspaper into small squares. Fill a cookpot three-fourths full. (It will get ink-blackened, but will scour out.) Add enough water to stir it easily. Soak overnight.

Cook over low heat, stirring occasionally. It will start to disintegrate into a mushy substance in about eight hours (less if under 4,000 ft. altitude). Cook one more hour, then cool. Squeeze pulp against a strainer to remove water. Then spread the pulp on several thicknesses of newspaper and let drain outside at least two hours. It can be left to drain overnight if desired.

Mix powdered wallpaper paste with water until it forms a thick mass. Add about one cup of this mixture to three cups of paper pulp, stirring and squeezing to eliminate lumps. This can be stored, tightly covered in the refrigerator, for weeks. Spread it with a knife or spatula onto your basket. As it dries, it clings to the basket like skin to a skeleton. This skin will not cover any flaws in the basket, so make sure your construction is aesthetically pleasing before you begin applying the pulp.

Color

I duplicate colors around me. Muted tones of chokecherries, sage, purple shadows, and the earth itself suggest colors.

I sometimes use Rit liquid dye, mixing it in the soaking vat before adding the willow and reed. I might mix acrylic paint or concentrated dye into a bowl of papier-mâché or pulp, or I might pour dye into the vat of pulp before couching. (In this instance wear rubber gloves.) Sometimes I paint a piece after it has been completed.

Embellishment

After the piece is dry, I often glue string, yarn, paper rush, and cut paper shapes onto it. Placement of bones, feathers, shells, bark, etc., should always be for a reason, and finding that reason is my biggest challenge. Control your urges to add too much.

An important part of embellishment, for me, is unseen, but I think many people have a sense of it. The parts make up the spirit of the whole. I often include a juniper berry, a poem, or just a group of words that speak of strength or wholeness right into the blender. It is macerated and applied to the basket.

I consider all of the materials I use as living, and I leave a hole somewhere in the papier-mâché application so my basket can be seen and can breathe its spirit out into the world. Sometimes, viewers have a sense of all this. Maybe they feel the "Basket Spirit."

Wano; reed, willow, shells, feathers, sagebrush, bark, and handmade paper.

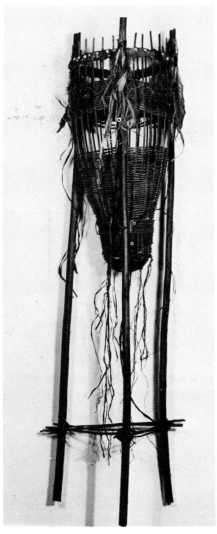

*Shaman Maund; reed, willow, bark,
feathers, shells, raffia, and paper pulp.*

Legend Jug; reed, raffia, cording, feathers, beads, and handmade paper.

The Basket Spirit

The great Spirit makes all things possible and is in every aspect of life. Gives life, substance, and our ability to create.

*"I am the tear of the little girl.
I am the wind that blows across the tree tops.
I am the grass that grows across the plains.
I am everything."**

*The Basket Spirit by Ken "Rainbow Cougar" Edwards, courtesy of Wintercount P.O. Box 576, Glenwood Springs, CO.

Pony Spirit; dyed reed twined basket covered with paper pulp, cording, paper rush, rawhide, and acrylic paint; photo: Carol Bakken.

Pierce the Cloud – Catch the Rain; willow twined with raffia and handmade paper.

Dhurrie Basket; 12" x 12" x 14" h; photo: Carol Bakken.

Spirit Shrine

Paper pottery embellished with furs and hides, bones and seed pods, rocks and metals to make an organic statement. Photo: Carlos Quintanilla.

ROSEMARY GONZALEZ

Embellished baskets.

The use of paper is cyclic in nature. With paper that is normally discarded, I shape my forms with the use of molds made in male-female fashion…a fusion of which is as ancient as the beginning of life itself. Old discarded paper, shredded and prepared, is "rebirthed" into new and different forms, then embellished with found objects such as wool, twigs, metals, feathers, threads, hair, and whatever else nature would like to provide. I enjoy sharp contrast—somewhat like the yin-yang idea: stark colors against bright bold colors, and straight lines accenting the curving line. I find great comfort in gathering these opposites and creating a balance between them.

I am influenced by the artifacts of the ancient cultures, and inspired by modern visionaries such as Brooke Medicine Eagle, Patricia Sun, and Eleanor Moore. I was especially influenced by the books written by Lynn Andrews titled *Medicine Woman, Flight of the Seventh Moon,* and *Jaguar Woman.* The need to create my olla-like forms came with the description and journey of the wedding basket that Lynn Andrews found to be the catalyst to her woman power and induction into an ancient sisterhood.

Paper Casting

To make cast paper vessels, one must first have an original form, such as a bowl or vase, that does not have any undercuts. (Undercuts will not allow the original form to come loose from the mold.) Plaster of Paris is used to make the mold from the original form. The molds are made to come apart so that the paper vessel can be removed in one piece. After the plaster mold is made, it must be thoroughly cured by drying in the sun or near a heater.

Paper pulp is made by blending chopped recyclable paper or by purchasing cotton linter or hemp. One may also want to try making paper from different kinds of plants that have been reduced down to pulp. This is a tedious procedure, but well worth the effort because the end product has such unique qualities. The paper pulp is placed by hand inside the molds with the pulp pressed firmly to the inside of the mold to ensure a smooth surface. After drying, the mold is taken apart and the cast paper vessel is removed. Acrylic varnish is brushed on the inside and outside of the paper vessel to protect it from the elements, make it waterproof, and to give sturdiness. At this point, the vessel is ready for embellishment. It can be painted, stenciled, dyed, woven over, or sewn into. Different objects can be adhered to the surface of the vessel. Cords can be wrapped around it. The surface can be batiked, and different types of paper forms can be glued to the surface to create a collage effect.

In creating this technique, I was motivated by its potential for durability, weightlessness, and shape. My first two loves are ceramics and paper, two extreme elements. Ceramics can hold dynamic shapes and can be touched and held, but are breakable and heavy. Paper is so versatile, lending itself to painting, drawing, dyeing, etc., and is light in weight. Its negative is that it does not hold up well when fondled or held for long periods of time. I needed the best of the two, so I combined them to make a dynamic, three-dimensional shape capable of being held, shipped, dropped without breaking, and versatile enough to keep my imagination going for a lifetime.

Vessels can be cast in many different shapes and sizes.

Making a Plaster Mold

1. Choose an object to be cast. Make sure this object does not have any undercuts that may be too complicated to cast. When casting a round rubber ball, for example, any point beyond the midline will make it impossible to remove the trapped ball from the hardened plaster.

2. Mix the plaster with water, following directions on package.

3. Spray the object to be cast with vegetable oil; then apply plaster to one side, making sure to have a supporting ring of clay edge so as to stop the plaster from running over to the other side of the object. Modeling and pottery clay both work.

4. Let the plaster set.

5. When the plaster has dried hard, cast the other side and let the plaster set.

6. When the plaster has dried hard, pull apart to free the original object.

7. Cure by allowing the plaster mold to dry in the sun or near a heater until it becomes light in weight.

Making the Mold

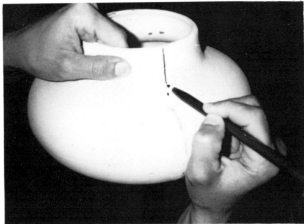

Making sure there are no undercuts, the original vessel is measured for its center and a line is drawn.

View of one side of the cast plaster mold on the original vessel. Photos: Javier Gonzalez.

This photo shows both sides of the mold. After the first side is dry, the second side can be cast over it without the two halves sticking together if the surface is sprayed with vegetable oil.

Casting the Paper

1. The slurry for casting should be thicker than for sheet making, and should be dipped from the vat with a strainer, and then be placed by hand into the mold and allowed to dry.

2. Open the plaster mold and remove the paper cast.

3. Waterproof the cast with varnish. The vessel is now ready for embellishment.

After the paper pulp is dry, the mold is gently pulled apart.

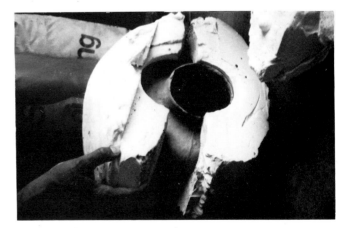

One side of the mold is taken off the vessel in order for it to dry further.

Paper vessel after it is taken out of the mold.

The artist applying paper pulp to the inside
of the closed mold.

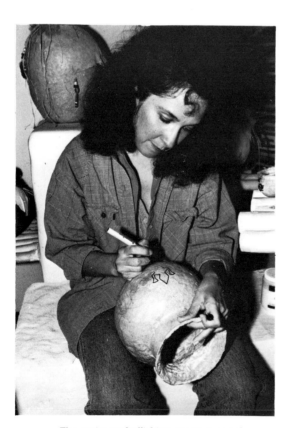

Paper vessel after embellishment.

The artist embellishing a paper vessel.

In the Beginning; 1987; philodendron paper with black and brown abaca paper, cast over gourd, and center woven paper with raffia; 8-1/2" x 8-1/2" x 11"; photo: Ruth Chambers.

BETZ SALMONT

Soft Rocks I & II; philodendron and abaca paper cast in plaster molds with dyed, coiled king palm seed stems; 7-1/2" x 8-1/2" x 11" & 7" x 7-1/2" x 8"; photo: Deborah Roundtree.

My first full-time job was as a papier-mâché artist in New York City. I worked in an old condemned warehouse on the East River, making wonderful animals and fanciful storybook figures for department store displays. Now, all these years later, I find myself returning to paper as a medium.

I have always liked working in three dimensions. After taking my first weekend workshop in papermaking, I couldn't wait to put this exciting medium to work somehow in basketry. I'd been working with gourds at the time and had a lot of them sitting around the yard. I decided to use these great natural shapes as molds to form my handmade paper. Working with paper is such a contrast to working with the leaves and seed stems I usually use in basketry. These basketry materials have a lot of influence on the finished piece. At times, they even try to take charge. Paper, on the other hand, is so submissive. I enjoy the feeling that I can be much more spontaneous with it.

Crecer; paper cast over gourd; 9-1/2" x 9-1/2" x 9-1/2"; photo: Ruth Chambers.

Crecer—side view; photo: Deborah Roundtree.

Detail of Crecer; photo: Deborah Roundtree.

Soft Rock; philodendron and abaca paper cast in plaster mold with dyed, coiled king palm seed stems; 7-1/2" x 8-1/2" x 11"; photo: Deborah Roundtree.

In the Beginning shown with the gourd it was cast over; photo: Ruth Chambers.

Emerge; philodendron and abaca paper cast over gourd; 8" x 10" x 12"; photo: Deborah Roundtree.

Confetti Cloud Sticks; dyed paper; 8" x 15".

Mosaic Cocoon with Cloud Sticks; dyed paper; 10" x 24".

CAROLYN DAHL

I make works of art to define my life through images, in the same way that I keep journals to clarify my thoughts through words. In each, I try to capture the awareness I experience in a life moment and to transform my responses into a visible marking, a recording, a clue to something that has passed.

Paper has, for centuries, been utilized as a recording medium, either by receiving words and images or by becoming the message itself. By constructing my three-dimensional forms of paper, I can take advantage of both means of communication. The design and colors convey my observations, my insights, and my emotional responses; the manipulated elements of torn, cut, twisted, and folded paper represent the kinetic energy and action of the creative process.

Mosaic Cloud Sticks; dyed paper; 8" x 15"; collection of Cia Hart, Houston, TX.

Lissa Hunter's Remember the Past Basket;
8-1/2"d x 11-1/2"h.

Alice Wand's Striped Box; 7" x 14".

Lissa Hunter's Confetti Basket; 4"d x 3"h.

Mary Merkel-Hess's Salvia; 10"h x 5" x 5".

Mary Merkle-Hess's Storm Box;
12"h x 5-1/2" x 3".

Alice Wand's basket with handle; 15"d.

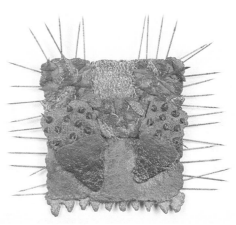

*Karron Nottingham Halverson's Splendor Active Principle;
23-1/2" x 25-1/2".*

*Mary Lee Fulkerson's Ghost Dance;
twined papier-mâché; 14" x 14" x 16".*

*Sue Smith's Summer Shadows; twined basket of handmade papers,
pine needles, and reed; 17"w x 13"h.*

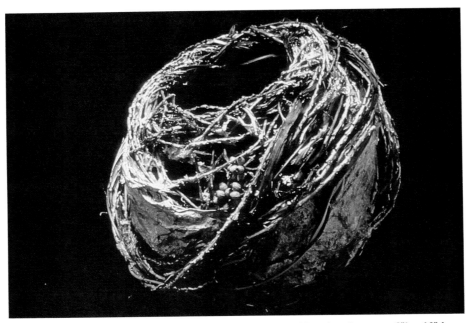

Sue Smith's Habitat; random weave with vine, bark, and handmade paper; 9"h x 10"d.

Donna Rhae Marder's Bowls; detail.

Donna Rhae Marder's Bowls; sewn maps;
collection of Rutgers University.

Rosemary Gonzalez's paper pottery; painted and
embellished with clay; photo: Carlos Quintanilla.

Rosemary Gonzalez's paper pottery; contemporary colors and motifs
give these pots a mystical effect; photo: Carlos Quintanilla.

Sharon Bock's Protected; 22"h x 24" x 21".

Ellen Clague's Not a Shred of Evidence to the Contrary; 6"w x 8"h.

Sharon Bock's Sweet Sweet Dreams; 14"h x 10" x 11".

Karron Nottingham Halverson's Aspiration Before Existence; 26" x 29".

Marilyn Wold; untitled; wauke handmade paper and
Dracaena draco weaving; 10"d x 8"h.

Carolyn Dahl's Red Lightning; dyed paper; 24"l x 9"h.

Carolyn Dahl's Cloud Stick Bowl; dyed paper; 15"l x 8"h.

Pam Barton's Kau' Desert; (from Volcano-scapes series).

Gammy Miller's Basket with Stamp Collection; coiled in waxed linen with mud, paper, and postage stamps; 2-1/4"h x 2-1/2"w.

Gammy Miller's Scrolls for Justice and Spring Rolls; coiled in wax linen with ash, mud, paper, and paint.

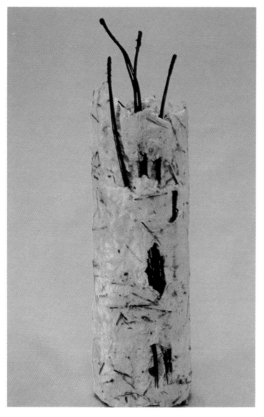

Judy Mulford's Spirit Basket Reach for the Clouds;
handmade paper and pine needles; 12-1/2" x 4".

Judy Mulford's Cloud Basket; cotton linter pulp;
6" x 8"h.

Betz Salmont; untitled; philodendron and abaca paper
cast over gourd; photo: Deborah Roundtree.

Betz Salmont's Paper Pouch for Gourd
Seeds; 8" x 8-1/2" x 8-1/2".

Molten Moon Vessel; dyed paper and silk; 7" x 17".

Pleated Rhythms; dyed paper; 11" x 17".

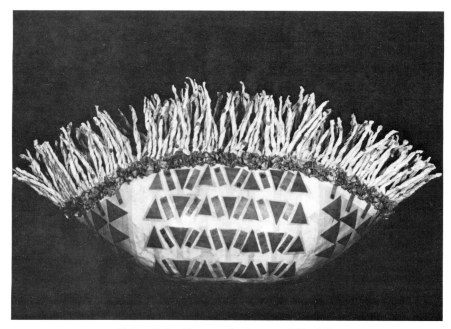

Yellow Solar Cocoon; dyed paper; 12" x 18".

Sky Catcher; dyed paper; 6" x 15".

Paper basket-bowl; handmade paper of wauke plant fiber; basket form woven with Dracaena draco plant fiber.

MARILYN WOLD

Plants…Fiber…Paper…An art? Yes. An exciting, wonderful one. When I first moved from the Pacific Northwest to the island of Hawaii, I was totally intrigued with the lush basketry material literally hanging from trees and growing everywhere. Baskets were just asking to be made. So I made them—plaited, twined, and coiled. Having studied papermaking in Oregon, I decided to also explore the Hawaiian plant fibers for papermaking. I turned to the history books to learn of the plants used by the Hawaiians for cordage, bark cloth, and basketry. The Polynesian people historically show skillful knowledge in their usage of tropical island flora. Some of the most refined, beautiful tapa and baskets of the past were produced here.

Wauke (*Broussonetia papyfirea*) was the plant fiber most used for tapa cloth. At one time, it was grown in groves so that it could be cared for and harvested for its bark. In many places it still grows wild and is the most beautiful plant fiber for handmade paper.

There is something so special about creating a basket or sheet of paper from a once once-living plant. The experience is one of reverence that brings you in touch with yourself and nature.

There are thousands of plants growing here on the island, and I have only just begun to tap my resources. The discoveries are exciting and rewarding, as each plant fiber produces a paper unique in color and texture. No two sheets are ever exactly alike; each has its own character and beauty.

Creating handmade paper continues to teach me. Its rewards are many. It has given me knowledge of botany and geography, a profound love of nature, patience, and a wealth of new friends.

Combining basketry and papermaking is a new and exciting experience for me. It is with great pleasure that I share my limited knowledge with you, in hopes that it will inspire you to create your own special techniques in both of these two wonderful art forms.

Handmade Paper of Hawaiian Fibers

After collecting and cutting a selected plant, the fiber is first cooked in a water and lye solution and then thoroughly washed. The fiber is further reduced by beating it. I use a heavy tapa beater with a large river rock as an anvil. A small amount of the beaten fiber is then mixed with water, processed in a blender, and poured into a large vat. A Japanese mucilage, *toro-aoi*, is added. This dispersing agent is the magic ingredient in the Oriental papermaking.

Now the sheets are ready to be formed. A polyester mold and deckle is dipped into the vat. The sheet is formed on the screen by rocking the pulp across the surface in a wave-like motion. The water is then drained through the newly formed sheet and laid or couched onto drying boards. After one day of drying, the sheets are peeled from the boards and stored for use.

Through many years of experimentation and research, I have developed this method of sheet formation, which is neither Oriental nor Western, but a blend of both.

Almost every plant that grows has fiber in it that will produce a sheet of paper. The reason some plants are preferred over others is the microscopic length of the cellulose fibers in that plant. Of the many kinds of plant fibers, the three I prefer to work with are bast (inner bark) fiber, leaf fiber, and grass fibers.

If you intend to do a lot of papermaking, I would suggest that you invest in some of the many wonderful books available on the subject. Like all art books, each one offers a little different information. I have included a list of reference books that I constantly use and could not do without. (See bibliography.)

Start a notebook. Each plant produces a different fiber and thus produces paper that differs in color and texture. Once a fiber is cooked and processed, it is hard to tell what plant it comes from, so try to work out a good labeling system.

A word of warning: *Know the plant you are working with. There are many toxic plants in the world.* Also, learn the botanical names of the plants. Chances are that if a plant proves to be a good one for papermaking, there will be others in the same botanical family that may also work well for you. Do not ignore the lowly weed. Some of the loveliest paper is made from milkweed.

Plants used for basketry will sometimes provide excellent fiber for papermaking too. Several times I have tried to break down a fiber for papermaking with no results, only to discover that it can be used as a strong basket fiber. The two work hand in hand.

Hawaiian Plant Fibers for Papermaking

Bast Fibers: These are the fibers found in the inner bark of plants. These fibers are the strongest and most desirable for fine paper. They also take more work because most of them need to be peeled or steamed from the woody core of the plant. They are then cut up, cooked, and beaten. Examples of bast fibers are wauke, hau, and banyan.

Leaf Fibers: These fibers are found in the leaves and stems of such plants as ginger and banana. They need to be cut up, cooked, and beaten.

Grass Fibers: The shortest and weakest of the three fibers, they need only be cooked and beaten. Examples of grass fibers are Job's tears, bulrush, and papyrus.

General Rules for Preparing Fiber

1. Select and identify the plant.

2. If possible, prepare while fresh. (The finest sheets of paper are made from fresh plants.)

3. Strip the bark if it comes off easily; cut into short pieces and put into a large cooking pot. Stainless steel is the most desirable.

4. If the fiber has been stored in the raw state and is dry, soak it overnight in warm water before cooking.

5. To avoid burns, always add cold water to the papermaking cook pot.

6. Carefully add measured lye or washing soda. One tablespoon of lye or soda to two quarts of water.

7. Bring the pot to a slow boil; turn the heat to low; cover and cook for two to three hours.

8. Please carefully read the safety precautions to protect eyes, hands, etc., while using lye or soda.

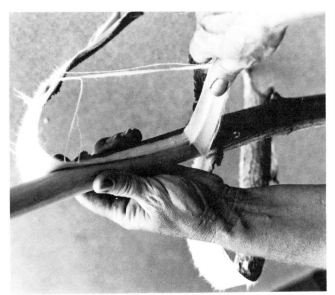

Wauke bast fiber is stripped from the woody core.

This is bast fiber, coiled. It can be dried and stored in this state until you are ready to process it.

9. Wear old clothes and rubber gloves. All fibers stain, and lye makes holes in anything, including you!

Storing Fiber

You can store the fiber at different stages, either directly after cutting it or after the fiber has been prepared and is ready for the vat. If you want to store it directly after cutting, simply dry it and then process at a later time. If you want to store fiber that you have already prepared, you can freeze or dry it. Freezing is better than drying because it is hard to get the water content back into the fiber after it has been dried. If you are working with fiber that has been dried, boil it and then let it soak overnight.

Be sure to label and date your fiber. Once it is frozen or dried, it all looks the same, and very much like anything else in the freezer. I have had lots of surprises.

General Rules for Washing Fiber

1. Wear rubber gloves and use cold running water.

2. A stainless steel colander or fiberglass window screen makes a good net to wash fiber.

3. All traces of lye must be washed out, so wash thoroughly.

4. As you wash, the outer bark and all the non-cellulose substances will wash away.

5. After thoroughly washing the fiber, it must be beaten to further break it down. After this step, it is ready to use or be stored.

6. To store fiber, label and date it. Then freeze or dry.

7. Good fiber feels silky after washing and looks like matted wool or hair.

Equipment for Papermaking

Molds: There are a variety of ready-made papermaking molds available for sale. However, you can very inexpensively make your own. Any size will do. I teach with an 8" x 10" frame. This is an excellent beginning size. Any wood frame is fine. It should be simple, and the wood should be smooth.

I use a polyester screen. You can use fiberglass window screen or polyester silk screen (the coarser, the better). If the material is woven too tightly, it will not allow the water to drain properly. You need good drainage.

Remember, the larger the frame, the harder it is to control the flow of fiber, so it's better to start out on a small scale.

Once you have your frame, staple the screen on tightly. Start by stapling the screen to the center of each side of the frame, then work your way around the frame, stapling on opposite sides and pulling the screen tight as you work. Finish at the corners, being careful to work out any wrinkles. I use silver duct tape to cover and protect the edges.

Deckle: A deckle is a frame that fits on top of your paper mold. It is the same size as the first frame. This helps to keep the pulp in place on the screen.

Most fibers must be beaten after cooking and washing. This creates fibrils on the surface of the cellulose fibers. I use my tapa beater and an anvil of river rock.

Vat and Drying Boards

Use a vat that is twice as large as your mold. This will allow good motion in forming your sheets of paper. I started out with a kitty litter pan, so use your imagination! Hardware stores sell a variety of plastic tubs. Keep in mind, though, that the larger the tub, the more fiber it takes to prepare the vat.

For drying boards, you want wood that is somewhat absorbent. I prefer to use untempered Masonite. Plywood or any smooth-textured wood would also be suitable.

When using plywood, use the smooth side. Plywood will have to be sanded periodically because the wet paper raises the grain.

Untempered Masonite board can be sprayed with any household wax when the sheets of paper no longer come off the board easily. Do not spray wax on plywood, however. The surface is too rough for the wax to do any good.

To prevent the boards from warping, be sure to stack them flat; also be sure they are dry before you store them.

Preparing the Vat

Fill the vat about one-third full of water. Now add a small amount of fiber to a blender with water. Blend and add to the vat. Continue this process until the vat is a little more than half full. This vat mixture is called slurry.

Next, add the magic ingredient neri, which is the Japanese word for dispersing agent. I use a powdered form that I order from Japan, but you can get the same thing from Lee Scott McDonald called PNS. The dispersing agent needs to be mixed a day or two ahead of time and stored in the refrigerator.

Paper can be made without neri, but the dispersing agent allows each fiber to be independent and not cling to others, therefore giving you a much smoother, more even sheet of paper.

Add a small amount of neri to the vat. It is very hard to say how much; you simply learn through experience. Too much and your sheet will not stay on the screen. Too little and the fiber will stick to the screen. Your water should feel silky. Add about one-fourth cup of neri to a dishpan-sized vat.

As soon as you feel it's right, you are ready to form your first sheet of paper.

General Rules for Vat and Sheet Formation

1. If the fiber has been stored, soak it overnight.

2. If the fiber is still long, cut it into pieces.

3. Put a small amount of fiber into a blender. Fill the blender about half full of water and blend. Pour into the vat.

4. Repeat the above process until the vat is half or more full. Add neri.

5. Stir vat until well mixed. Now you are ready to form your first sheet.

6. Be sure the screens are clean. Rinse them in cold water before using.

7. The sheets are formed with a rocking motion. Usually two dips make a nice sheet. Experimentation is the key.

8. Drain well, then turn out onto drying board.

9. Remember, the sheet is very fragile—treat it gently.

10. Blot out liquid from back of screen with sponge.

11. Lift mold and lay towel over newly formed sheet and roll out more moisture with a rolling pin.

This is wild ginger. The stem of the plant is the part cooked for paper fiber. Strip all the leaves and cut stem into short sections.

Always use COLD water in your cooking kettle. After adding water, add one tablespoon of lye for every two quarts of water.

Removing and Storing Finished Sheets

To remove sheets from the drying boards, I use a table knife to lift one corner. They usually come off easily. Be sure the sheets are good and dry. I usually leave my sheets on the boards overnight and remove them the next morning. Once removed, they are stacked on a piece of clean, flat cloth with a second piece of cloth placed on top.

The sheets need to remain under pressure and flat for a few days. I find that a weight placed on top is helpful.

Paper should be cured. I try to cure mine a month or more before I use it for artwork. Because this paper is unsized (waterleaf paper), it can be sprayed with a fixative before using for artwork.

You will notice that there is a right and wrong side to your paper. The smooth side is considered the right side. This is the side that was next to the drying board. The rougher side is the one that was next to the screen.

To protect the paper from insects, I use pure camphor in the drawers that I store it in. Be sure to store your sheets flat in a dry place.

The Use of Handmade Paper in Basketry

There are many ways to use your sheets of paper in basketry. Be creative. The three that I prefer and use the most are: rolled or spun paper; sheets applied to basket frames; and pulp applied onto basket form.

Rolled or Spun Paper

In Japan spun paper is called shifu. It can be made as fine as the thinnest yarns. This is woven into cloth that is used for kimonos, obi sashes, underclothing, purses, ties, and many other articles.

For my purposes, I roll and spin the paper by hand, but not nearly as fine and beautiful as the Japanese shifu. You can use almost any paper; even wrapping paper or paper bags will do.

To start, determine the grain of the paper. This can be done by tearing a small piece. The grain is the direction in which it will tear easily. This is the direction you want to make the cuts. Cut the sheet into narrow widths with an X-acto knife. If you want one continuous string, you must cut it as the shifu makers do. This is done by folding the sheet four

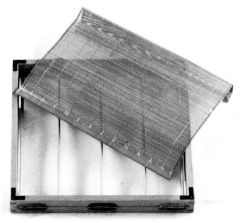

Japanese su and keta.

Simple Japanese-type frame on the left. Western-style frame and deckle on the right.

times, but leaving a margin on both sides. In other words, you do not cut through the edge of the sheet. Next, tear every other side of the margin, which will give you a long, continuous strip. Using a concrete block to roll it on, dampen your hands and the top of the block. Roll the strip very gently on the surface of the block. Once you have your paper string rolled, let it dry thoroughly and use it as a weaver or in a coiled basket.

Sheets Applied to Basket Frame

After your basket is formed or partly woven, you can cover the structure with dampened sheets of handmade paper. The sheet is carefully dampened with a sponge or placed briefly between wet cloths. Then it is gently draped over the basket frame and molded to the structure. Once it has dried, I apply a thin coat of rabbit-skin glue. After you have let all of this dry completely, a coat of matte varnish can be applied to protect the finish of the paper.

Paper Pulp Applied to Basket Forms

This method is pure fun, with surprising results. Prepare your paper pulp and vat just as you would if you were going to form sheets of paper. I usually make paper first, and then at the end of the day when the vat is rather thin, I cover a basket or a form with leftover pulp. You can either dip the pulp up and pour it over the form, or you can dip the whole basket into the vat. Be sure the basket is dry and clean.

After you have built up the thickness you desire, let the vessel dry thoroughly. Usually I will let it dry for a week before I add any embellishments. You may also want to add more weaving. There are many directions you can go with this. Adding feathers, beads, or whatever you would like is fun and creative.

Rocking motion of the screen.

The sheet of paper is formed when the liquid pulp is rocked back and forth on the surface of the screen.

Forming the Paper

Mixing the neri.

When you are satisfied with the evenness of the sheet, let it drain well before couching it onto the drying board.

It is not always necessary to use the deckle. Here I am forming a very thin sheet. If I desire a thicker sheet, I use the deckle on top of the frame.

Forming the Paper

Carefully sponge the liquid from the back of the screen. This is done with a downward pressure.

Forming the Paper

Do not move the sponge back and forth because it will also move the fiber under the screen. This is called bruising and causes thin spots in the paper.

Be sure that the perimeter of the sheet has made contact with the drying board. Carefully, with a quick, even motion, remove the screen.

Place a dry towel over the wet sheet and use a wooden roller to gently roll it in both directions. It is very fragile at this point, so don't use too much pressure. As soon as your drying board is full, set it somewhere out of the way so that your newly formed sheets can dry.

Rolled or Spun Paper

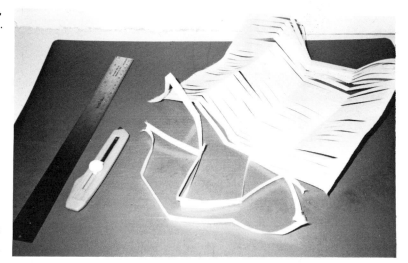

Paper cut to form continuous string.

Rolling paper string on a concrete block.

Starting a coiled basket with paper string.

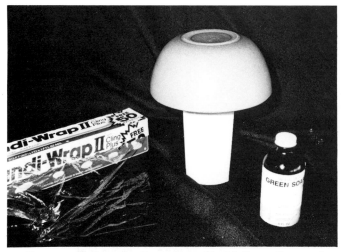

Choose a simple mold. Cover the mold with Green Soap and then tightly wrap it in plastic wrap.

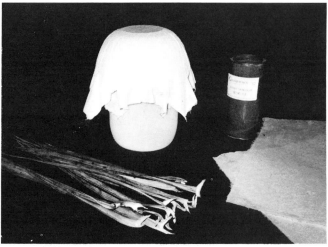

Dampen two sheets of paper and carefully drape the form. Allow to dry thoroughly. Apply a thin coat of hot rabbit-skin glue; one tablespoon of dry granules to three cups of hot water.

When the glue is completely dry, you may add fiber woven or laid on in a pattern.

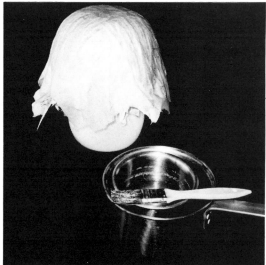

Dampen and mold on two more sheets of paper. Dry again and coat with another thin layer of hot rabbit-skin glue.

When this has dried, remove from the mold and apply medium to the inside and outside of the form. You may want to give the whole piece a second coat of acrylic medium.

Finished bowl-basket.

Basket mask–hau, Dracaena draco, philodendron sheath, and coconut fiber; covered with paper pulp made of wauke fiber.

Paper basket-bowl; handmade paper of wauke plant fiber; basket form woven with Dracaena draco plant fiber.

Basket-bowl; handmade paper, reed, and paper strings.

Wauke (Brousontia papyfifera).

Common Name: Balloon Plant

Botanical Name: Gomphocarpus physocarpus

Where It's Found: Vacant lots; roadsides.

When To Harvest: When plant is green and blooming.

How To Collect: Cut whole plant at base of growth.

Part To Use: Stem

Preparation: Discard seed pods, flowers, and leaves; cut stems into pot-size pieces and cook in lye water for two to four hours.

Miscellaneous: This plant is native to South Africa.

Common Name: Beach hibiscus tree

Botanical Name: Hibiscus tiliaceus

Where It's Found: Grows at sea level in Hawaii.

When To Harvest: Anytime

How To Collect: Cut long slender new growth or slender limbs.

Part To Use: Inner bark

Preparation: Strip bark off from woody core; soak in salt or fresh water, changing the water each day for one to two weeks to allow the layers of fiber to separate; rinse and dry before storing.

Miscellaneous: This fiber is one of the strongest in all the Polynesian islands. It is used for weaving, cordage, net making, etc.

Common Name: Chinese banana

Botanical Name: Musa nana

Where It's Found: Good at sea level; common in yards.

When To Harvest: Anytime

How To Collect: When bananas have ripened, the tree can be cut down.

Part To Use: Trunk

Preparation: Peel off dark outer bark; cut trunk into pieces and cook in lye two or more hours; wash and pound.

Miscellaneous: The paper color is in the light brown tones. The paper is very strong and lovely to work with.

Common Name: Chinese Banyan

Botanical Name: Fiscus retusa

Where It's Found: In yards or parks.

When To Harvest: Anytime

How To Collect: Cut small slender branches.

Part To Use: Limbs, small branches.

Preparation: Cut into pot-size lengths; discard leaves and fruit; cook in lye water for two to four hours; rinse and remove woody core.

Miscellaneous: There are hundreds of species of fig family in the Hawaiian islands; this species is a large evergreen tree with slender hanging aerial roots.

Common Name: Common banana, apple banana

Botanical Name: Musa paradisiaca

Where It's Found: Most any elevation; this banana tree grows everywhere.

When To Harvest: When a hand of bananas is ready to be cut, the banana tree is cut down so a new one can grow.

How To Collect: When the fruit is removed, take the main part of the tree trunk.

Part To Use: Main trunk

Preparation: Peel outer coarse bark; cut trunk into pieces and cook in lye two or more hours; pound, freeze, or dry.

Miscellaneous: Paper color is in the darker brown tones; this banana tree is very tall and grows everywhere; the paper is very strong and has a wonderful texture.

- -

Common Name: Great Bulrush

Botanical Name: Scirpus validus

Where It's Found: Grows on the edges of fresh or brackish water marshes.

When To Harvest: Anytime

How To Collect: Cut plant at waterline.

Part To Use: Stem

Preparation: Cut stems into pot-size lengths; boil in lye water two to three hours.

Miscellaneous: Formerly used for thatching or plaiting.

- -

Common Name: Heliconia

Botanical Name: Heliconia

Where It's Found: Yard plant

When To Harvest: Anytime

How To Collect: Cut plant at base of stem.

Part To Use: Only stems are used. (If plant is young, leaves can also be used.)

Preparation: After cutting, cook in lye for two hours or more; wash, pound, and dry or freeze.

Miscellaneous: The paper color is yellow; this plant is a native of South America; all species are good.

- -

Common Name: Hibiscus

Botanical Name: Malvaceae

Where It's Found: Over 1,000 species; grows well at all levels.

When To Harvest: Anytime

How To Collect: Cut small branches or new saplings.

Part To Use: Small branches, saplings. (Note: the saplings are easy to harvest and it does not harm the tree.)

Preparation: After cutting, strip or cook with stem on in lye two to four hours; wash and pound.

Miscellaneous: The paper colors are shades of tan and yellow.

- -

Common Name: Ilima

Botanical Name: Sida cordifolia

Where It's Found: Vacant lots; along roadsides.

When To Harvest: Easier to work if harvested after the rainy season.

How To Collect: Cut whole plant from base.

Part To Use: All but leaves and flowers.

Preparation: Cut branches into pot-size lengths; can be peeled from woody core or cooked as is; cook in lye water two to four hours; rinse and peel, discarding woody core.

Miscellaneous: The paper is pale yellow.

Common Name: Job's tears
Botanical Name: Coix-lacryma jobi
Where It's Found: Damp wasteland along drainage ditches.
When To Harvest: Young green growth will yield white paper; old dried cuttings yield tan paper.
How To Collect: Cut stalks at base of plant.
Part To Use: Use only the stalk of the plant; discard the flowering head.
Preparation: After cutting, cook in lye and wash; does not require beating.
Miscellaneous: Paper color is creamy white; native to East Indies; first collected in Hawaii in 1895; seeds used in jewelry and leis.

- -

Common Name: Mamaki
Botanical Name: Pipturus albridus
Where It's Found: In the forest mostly between 1500-4000 feet altitude.
When To Harvest: Anytime
How To Collect: Cut small branches.
Part To Use: Small branches
Preparation: It can be steamed or cooked with stem on in lye two to four hours; wash and pound; freeze or dry.
Miscellaneous: The paper color is pinkish tan; also used for tapa cloth.

- -

Common Name: Melochia
Botanical Name: Melochia umbellata
Where It's Found: Found at sea level, mostly Hilo side.
When To Harvest: After rainy season
How To Collect: Cut slender branches of new growth.
Part To Use: Branches
Preparation: After cutting, cook in lye two or more hours; wash and pound.
Miscellaneous: The paper color is light pinkish; this plant is native to India, brought to Hawaii for reforestation. It yields a strong dye, dark red in color.

- -

Common Name: Milo
Botanical Name: Thespesia populnea
Where It's Found: Low elevation, along coastline.
When To Harvest: After the rainy season
How To Collect: Cut small branches or new growth.
Part To Use: Bark of slender branches; discard leaves.
Preparation: Cook in lye.
Miscellaneous: The paper color is yellow tan; this plant is hard to break down, much like hau; the milo wood is favored by Hawaiian bowl carvers.

- -

Common Name: Mulberry, black mulberry
Botanical Name: Morus nigra
Where It's Found: Wild along the roadside; in yards.
When To Harvest: Harvest after the rainy season.
How To Collect: Cut new growth or slender branches.
Part To Use: Stems; discard leaves.
Preparation: After cutting, cook or steam, then cook in lye two or more hours; wash and pound.
Miscellaneous: Paper color is white. It is a very strong fiber and the paper is lovely.

Common Name: *Paper mulberry*

Botanical Name: *Broussonetia papyrifera*

Where It's Found: *Grows in the higher, wetter elevations.*

When To Harvest: *Harvest after the rainy season.*

How To Collect: *Cut slender branches of new growth from trees.*

Part To Use: *All limbs*

Preparation: *After cutting, cook or steam; then cook in lye three or more hours; wash and pound; then dry or freeze.*

Miscellaneous: *Paper color is white to light tan; this is the fiber used for tapa or kapa cloth; it is very strong.*

Common Name: *Papyrus*

Botanical Name: *Cyperus papyrus*

Where It's Found: *Watery swampy places*

When To Harvest: *Anytime*

How To Collect: *Cut stems of plant at base or waterline; leave on some of the red sheath at the base of the stems.*

Part To Use: *The stems are best, but I have also used the whole plant.*

Preparation: *Cut stems of plant; cook in lye three to four hours; wash, pound, and cut; freeze or dry; be sure to rinse in a screen, as the fibers are small.*

Miscellaneous: *The paper colors are pale, rich browns and golds.*

Common Name: *White Ginger*

Botanical Name: *Hedychium coronarium*

Where It's Found: *Yards, or along roadsides.*

When To Harvest: *Anytime*

How To Collect: *Cut new growth at base of plant.*

Part To Use: *Stems*

Preparation: *Cut stems into pot-size lengths and cook in lye water two to four hours.*

Miscellaneous: *The flowers are a source of commercially produced perfume.*

Split Column; handmade paper, organic material, thorns, and excelsior; 32" x 10" x 8".

SHARON BOCK

I live in a rural setting and am constantly aware of nature's colors, materials, and variations. I use my observations of nature, along with the inherent characteristics of paper, to express my interest in internal and external space. Paper, as well as nature, can have an unlimited range of tactile qualities. It can take on various shapes and show an infinite variety of colors. Handmade paper, to me, is an integral part of all my work—not an addition or decorative element—and it provides an immediate expression of my ideas. I divide my various types of work into three areas, each one using paper a little differently. I use paper as an internal or external skin on my coiled baskets; with reed and wild material on twined baskets; and by making forms totally of paper, either cast or on a reed framework.

The technical approach I use to make paper for my baskets and three-dimensional forms is very simple and straightforward. To begin, I choose cotton or abaca linters for my pulp. (I have made pulp directly from organic material such as iris or cattail leaves, but I prefer the abundance and availability of purchased linters over the collected organic material.) After pulping my linters, I dye them the color I want. I create my own color palette by dying the linters primary colors and using a blender to mix shades or secondary colors. I have used various dyes, but find Procion M to be the best. It is colorfast and does not bleed when used correctly. My next step is to fill the vats or containers with pulp. I mix the appropriate amount of pulp, making it the correct color or shade.

At this point, I can add extra material to the vat as a design element. I am very free with the selection of this extra material, which can be organic or man-made. I use horse-hair, sisal, seeds, leaves, threads, glitter, or whatever I feel will enhance or carry out my design. If I choose not to add extra material to the vat, I am ready to prepare my work

Ripe Vessel II; handmade paper, reed, organic material, and gauze; 21" x 40" x 30".

area. I use two layers of towels, with a final layer of white sheeting to cover the entire area that will be accepting the wet pulp. With my work area prepared and the color, texture, and density of pulp correct, I am ready to lift the pulp out of the vat. I do this using a screen without a deckle. After couching the paper on my work area, I don't press it, but let it become workable to the touch. When the paper is ready, I place it directly on the coils, reed, or framework, overlapping each piece. If at all possible, I finish the work in one time period because the wet paper layers bond to each other. I do use methyl cellulose to bond one layer of paper or one area to another if necessary. On large pieces I use gauze between the layers of paper to give the form added strength.

I finish the piece by making it waterproof. To accomplish this, I wait until the work is completely dry, then spray or paint sizing on the paper. The sizing takes about a week to completely cure. I have found the paper on my baskets and paper forms to be extremely sturdy after the finishing and waterproofing. They take the wear and tear of shipping to and from shows, and the rigors of exhibitions, and return in excellent condition.

The photographs on page 113 and 114 help to explain how I add extra material to paper pulp. After the first steps of pulping and dying the linters to the desired shade, I decide whether I will add any extra material to the pulp. Some pieces are very well suited to smooth, plain paper, while others have a need for strong texture or a varied surface quality. I often use contrasting surface textures as part of my design, with a smooth paper skin on either the inside or outside of the piece and a visually textured skin opposing it.

I have experimented with a number of additives, and chose those that I felt would last and enhance the paper quality. Organic additives such as leaves, seeds, petals, and roots work best if they have been dried. Green, fresh material can decompose and cause mold in the paper. Dried organic material, however, will last and not harm the paper. The amount of material I use depends on my design and how much variation I want to show on the surface of the final piece. If the material I choose is in large pieces, I tear or cut it into the size I feel will carry out my idea. Then I put the material directly into the vat of wet paper pulp, stirring well. If I choose sisal, thread, horsehair, or ribbon as an additive, I cut it to the desired length and add it in the same manner, stirring well. Some of the chosen material might float on top of the pulp, while other material will sink to the bottom. Occasional stirring and the addition of new material will solve these problems. During this process I frequently pull samples of the mixture to test the appearance of the paper.

Paper alone has strength when placed in layers and dried. I have found that some of my pieces need extra strength because of their shape or because they need durability for shipping and handling. To achieve this extra strength, I place gauze or cheesecloth between the layers of paper. I purchase these materials in bulk quantities, although smaller amounts can be found in fabric shops and variety stores.

Before placing gauze or cheesecloth on the paper, I apply methyl cellulose as a bonding agent. I first brush the methyl cellulose on the paper, then lay the gauze directly on the coated paper layer. Before I add any additional paper, I repeat the application of methyl cellulose to assure a strong bond between the gauze and the new paper layer. The number of layers depends on personal preference, the piece, or the area to be strengthened. When the piece is totally dry, the layers of paper and gauze will adhere to one another and give the work the desired strength.

Creativity grows with experimentation. The list of man-made and organic materials that can be added to paper is limited only by one's choices and preference.

*Red Lipped Basket; coiling with jute, handmade iris leaf paper, thread, and reed;
13" x 26" x 18".*

*Night Dreams II; handmade paper and tar paper;
26" x 18" x 20".*

*Spring Vessel; handmade paper, organic material,
and gauze; 21" x 11" x 21".*

Preparing the Materials

Materials to be added to the pulp.

Cut the materials before adding them to the pulp.

Add the material to the paper pulp.

Adding Materials to Pulp

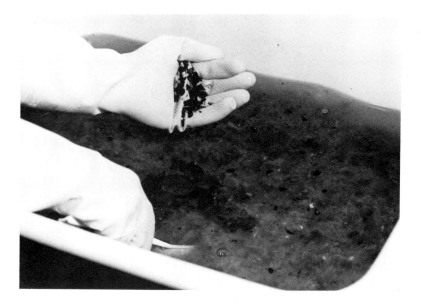

Additional material can be added.

Stir the pulp and additives.

Pull paper pulp for the finished piece.

Applying Gauze

Apply methyl cellulose to the first layer of paper.

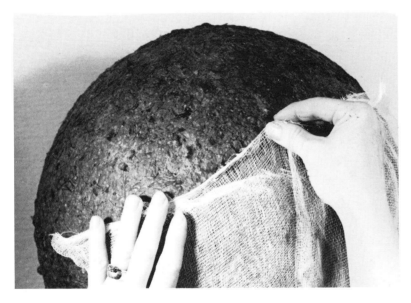

Add gauze to the paper layer, directly onto the methyl cellulose.

Apply another coat of methyl cellulose to the gauze before adding the next layer of paper.

Ripe Vessel I; handmade paper, organic material, horse hair, and gauze; 20" x 13" x 21".

Protected; handmade paper, reed, organic material, and gauze; 21" x 22" x 20".

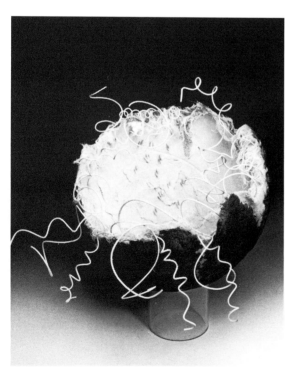

In the Beginning; handmade paper, reed, and gauze; 17" x 15" x 20".

Sweet Sweet Dreams; coiling with bailing twine, handmade paper, wire, thread, reed, and glitter; 14" x 10" x 11".

and copper wire; 23" x 14" x 13".

linen, and bamboo skewers; 32" x 12" x 14".

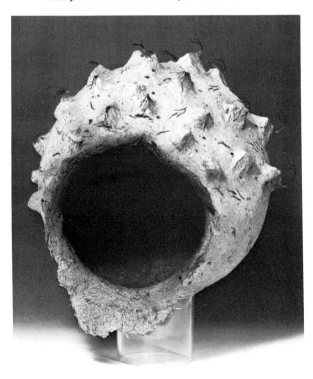

Secret Place; handmade paper, thread, wire mylar,
cocoon, and bamboo skewers; 9" x 8" x 8".

Ripened; handmade paper, wire, organic material,
air brushed surface design; 16" x 18" x 17".

Friendship Bowl; coiling with yarn, handmade paper, organic material, reed, and printing; 5" x 16".

Baskets with Beaver Sticks; handmade paper, reed, organic material, beaver sticks, and waxed linen; 30" x 11" x 10".

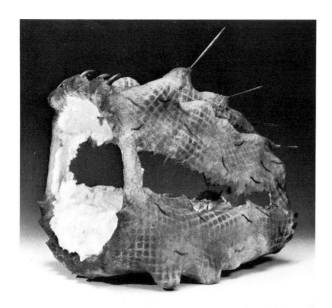

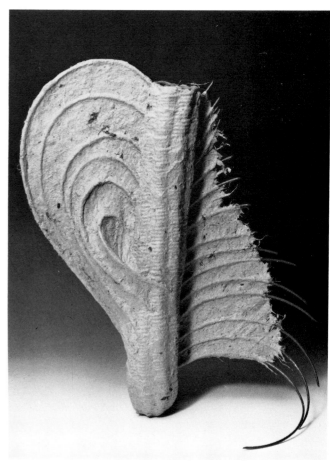

Spirit Carrier; handmade paper, reed, thorns, turtle toenails, and capacitors; 10" x 12" x 10".

Ribbed Basket; handmade paper, reed, and organic material; 21" x 16" x 4".

Blue Iris; 1986; paper construction; 8"h x 6" x 6".

MARY MERKEL-HESS

Branches; 1987; molded paper; 8"h x 14" x 14".

My involvement with paper as a three-dimensional medium began when I made a few molded paper baskets. After years of working in metal, the freedom the paper offered me was heady, and I began to explore the possibilities of paper. I still make molded baskets using layers of thin paper, but I also construct vessels using heavy paper boards with a collage overlay. I use different types of papers, only occasionally making my own, and I often include my own drawings in the collage, or paint directly on the surface of my vessels.

An interest in historic vessels and containers is my inspiration for basketmaking. I combine a love for strong, simple shapes with an appreciation for natural phenomena—grasses, flowers, small animals. My vessels recall our ancient tradition of vessel-making, as well as being a visual record of my response to the natural world.

*Storm Box; 1987; paper construction;
12"h x 5-1/2" x 3".*

*White Bird; 1986; paper construction;
18" x 5" x 3-1/2".*

Iris; 1985; paper construction; 12"h x 8" x 8".

Pink Square; 1987; molded paper; 6"h x 8" x 8".

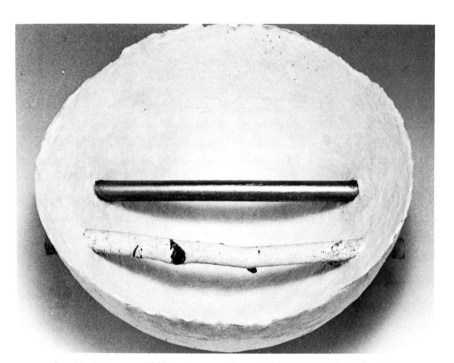

Double Exposure; 1985; paper, wood, and white birch; 10" x 14" x 14".

Untitled Black Basket; 1985; molded paper and wood; 8"h x 14" x 14".

Small Moons; 1987; paper construction; 18"h x 12-1/2" x 3-1/2".

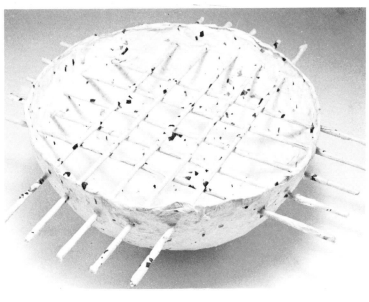

Grid Basket II; 1986; molded paper; 8" h x 18" x 18".

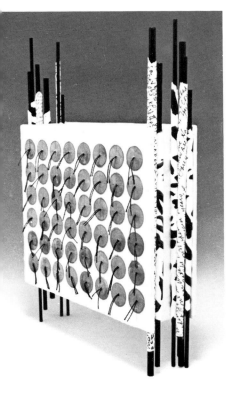

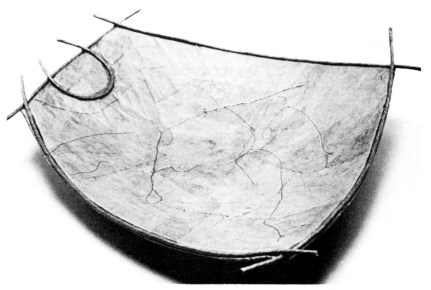

Feeding the Sun With Blood; 1986; paper construction; 17"h x 11" x 3".

Small House; 1987; molded paper; 2"h x 12" x 7".

Circle in the Square Basket; 5-1/2"h x 11-1/2"d; photo: Stretch Tuemmler.

LISSA HUNTER

Making baskets came to me in an odd way. Ten years ago, I was about to have surgery which I knew would surely keep me incapacitated for several weeks. At the very least, weaving at my floor loom seemed physically unlikely, and so a friend suggested that I try something a little less athletic and gave me a copy of Dona Meilach's book *A Modern Approach to Basketry.*

Armed with some bits of handspun, vegetable-dyed wool yarn and a bundle of sash cord, I followed the pictures and coiled a basket of dubious merit. I was sold on this intimate, friendly process of making a vessel form.

The genesis of my papermaking is not so easy to pinpoint. Around 1980 you couldn't pick up a craft or art magazine without seeing handmade paper in some form or another. The common and ubiquitous medium was finding new glamour and focused attention. Not being one to miss a bandwagon, I wrote to the Government Printing Office for a flier on papermaking. The basic materials and equipment it required were toilet paper and a blender. And I haven't progressed much beyond that, as a matter of fact.

The first sheets of paper were uneven and without character, but still magical. To watch the formation of so common a material is fascinating; a mystery solved. Then I started throwing in everything the kitchen or studio had to offer. Some sheets had so much oatmeal and parsley in them they were absolutely edible. Some could qualify as felted fabric, with scraps of thread thick enough to resemble placemats. But after the initial hysteria of discovery diminished, more structurally sound and less perishable paper was produced with reasonable efficiency, even with primitive equipment. And a second interest was born.

Bead Basket; 9"h x 9"d; photo: Craig Blouin.

Of course, at this point, the question became how to use these two media together. How could I integrate paper and basketmaking into a personal visual statement? At first it seemed impossible without imposing artificial constraints on each. After all, baskets are three-dimensional and paper is two-dimensional. What they have in common, I reasoned, is surface. Each has a surface. By superimposing one over the other, paper on basket, I could develop the two-dimensional possibilities of the basket, and the three-dimensional possibilities of the paper.

Another point in common was the similarity of the qualities in the materials. The baskets I was making at the time were coiled of raffia over paper cord, and the paper I was making was of abaca fiber. I was struck by the "paperiness" of raffia and the "grassiness" of abaca paper.

At some point I became concerned that I was betraying a tradition I respect and admire, the tradition of the basketry form. One of the engaging qualities of baskets is the relationship of the physical structure and the visual effect of that structure. One can see the structure immediately, and the structure dictates the form of the basket.

Thatch Basket; 15"h x 8"d; photo: Kirby Pilcher.

What you see is what you get. I felt that by covering the structure, I would be denying one of the qualities I like so much about baskets, and that I might be forcing these materials only for the sake of novelty. But as I began experimenting, I realized that there was enough of a physical and aesthetic affinity to make this work. Because the paper is thin, it still reveals some of the basketry form beneath. And of course the structure is apparent on the inside.

I like to think that these baskets are extending the basketry tradition, not denying it. All forms that we consider traditional today must have seemed radical at some point. I can just hear it: "Look, pal, you can't split that ash log and weave those strips into a basket. It will never work. Vines have worked this long and will continue to work. What you're making isn't even a basket!"

In extending tradition, I think we bear the responsibility of guarding and preserving the essence of that tradition, of respecting the qualities that have endured. Integrity of structure and attention to detail, both structurally and visually, must be maintained. The other side of this coin is that anything is possible as long as we maintain that integrity.

Those who coil baskets must acknowledge the presence of American Indian design in their work. Tribal basketry is so rich and complex that I doubt there are patterns or forms not already produced. But because I work on the surface, unrelated to the structure of coiling, other inspirations and influences can be explored.

Of course I look at ethnic coiled baskets. And I also look at pottery forms. Coiled pottery is especially close to coiled basketry, and so the forms can relate directly. Even wheel-thrown pottery has the same inherent symmetry.

The patterns of quiltmaking and stitchery also offer a rich resource for surface design. Because I often work with pieces of fabric, leather, and paper in a pattern, I can draw upon the logic of pieced quilt construction. And stitchery creates not only a visual element but a texture as well.

Ethnic finishing techniques used in weaving and garment construction can become the finishing touch on a basket, the little extra something that sets it apart. I often use braids, tassels, edging, and knotting that I have seen in South American and African textiles.

Inspiration can come from the strangest places. One day I was shopping at a fruit and vegetable market nearby, when I was struck by the perfect elegance of French pears,

Sundown Basket; 7"h x 9"l x 6"w; photo: Craig Blouin.

What Goes Around Comes Around Basket;
4"h x 7-1/4"d; photo: Stretch Tuemmler.

lined up row by row with their stems aligned, in a pine crate. They were that lovely spring gray-green color, with dark stems, and each one tipped with a healthy blop of red sealing wax, like so many berets. On the way home, I stopped at store after store looking for sealing wax, and finally found some in a stationery shop. The result of this inspiration was *Saddle Basket* with wax-tipped fringes.

The U.S. Patent Office gave me the idea for *Thatch Basket*. While looking up textiles in the main catalog, I ran across thatching. Intrigued, I went into the stacks and pulled the dozen or so entries from the appointed shelf, and was amazed by the intricacy of thatched roofs and outbuildings. My basket uses none of the patented methods, but it does use the idea of thatching.

Most of all, the nature of the materials I use inspires me. There are many differences between paper surfaces—whether they are torn, cut, folded, painted, or pierced—and these qualities demand certain treatment. They speak to me and tell me what to do. By working with other materials as well, I not only expand the variety of qualities available to me, but also the juxtaposition of those qualities. Torn paper edges look like the natural edges of leather, but cut shiny paper relates more closely to metal foil. I can look for similarities or differences in materials and work the design from there.

I have found that there are three basic elements in approaching any design problem: (1) materials, (2) structural techniques, and (3) imagery, or what the artist wants to say. I can start at any of the three points and select the other two accordingly. If I find a material that excites me, I must then find a technique that will use the innate properties of that material which will produce a certain image or statement. If I know that I want to use a certain construction technique, I must find the most suitable materials to support the structure. If I begin with the image, with something I want to say, the materials and techniques chosen must produce that desired image. The more techniques and materials I have to choose from, the more I can say.

It is rare in this life to feel that we are wholly in charge. At least with my baskets, I feel I can make choices. I can manipulate the materials and produce an object which is an extension of my vision. Surprises do happen: most good, some bad, all workable.

Long Ago and Far Away Basket; 4-1/2"h x 8-1/4"d;
photo: Stretch Tuemmler.

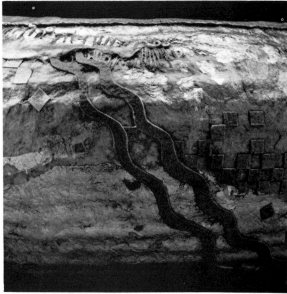

Long Ago and Far Away Basket, detail;
photo: Stretch Tuemmler.

Papermaking

I am not a papermaker. I make some of the paper I use for my work, but I make only the type I need, and only as much as I need. I use primitive equipment and prepared pulp. As much as I admire the processes involved in papermaking, I must admit that I want the paper I need without the tremendous amount of effort it takes to make a full range of textures, colors, and fibers.

The paper I do make is essential to what I do. It is soft, textured, and used for the skin of the basket, which is then embellished. I can count on it because I make it the same way each time.

I purchase other papers for the embellishment. Some are handmade, and some are commercially made 100% rag or acid-free papers. The number of small handmade paper concerns is growing, and each seems to have a personality all its own. This makes it a treat to do business. Sometimes you might not receive quite what you expected if you order by mail, which is what I usually do, but as long as the quality is there, those surprises can work for you. They can lead you in new directions.

I use what might be called the kitchen method of making paper, with everything centered around the sink—blender and all. This tends to disrupt other kitchen activities, but it is the most efficient place to work.

The equipment I use is basic and easily collected. I bought a blender at a yard sale for five dollars. The more powerful the blender, the better the job it will do. I use it only for hydrating paper pulp, not for food preparation. That way, I don't have to worry about food in my paper or paper in my food. A general utility plastic bucket serves well to soak the pulp in preparation for hydrating.

I now use a commercially available mold and deckle, but when I first started I used a piece of metal window screen mesh. It is sold by the foot at hardware stores and costs very little. It was fine for a small number of sheets, and in fact created a nice surface texture because of the woven pattern of the wire mesh. But as I began making more and more paper, I found its floppiness inefficient and irritating, so I held a cake cooling rack under it to provide support. This was a little better, but still not efficient for large runs. That is when I decided to buy a mold and deckle. They can be quite expensive, so I settled for a simple mold which has a heat-shrunk polymer mesh. It has been quite serviceable, and its 8-1/2" x 11" sheet size is adequate since I cut or tear the paper into smaller pieces anyway.

For a pulp vat, I use a utility tub from the hardware store. It is molded plastic and about 24" x 18" x 6" in size. A little deeper would be better, but this one is adequate. I started out with a disposable aluminum roasting pan, and I ended up with more water and paper on the counter and floor than in the pan.

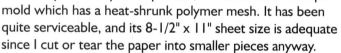

A large, clean all-purpose sponge and a hard, clean flat surface complete the list of needed equipment. I used to use a 4' x 4' sheet of Masonite, but eventually the fibers started being picked up in the paper, so I constructed a Formica-topped table. A friend uses an old dinette tabletop and it's terrific. The advantage of a movable piece is that it can be removed to help the paper dry quickly and stored out of the way when not in use. I use the table for other work as well, so it is quite valuable. I don't use blotters or felts but they can be useful.

Pocket Stilt Basket; 15"h x 6"w; photo: Color Marketing Concepts.

Papermaking Equipment

- Blender
- Bucket
- Mold and Deckle
- Pan
- Sponge
- Hard, Flat Surface

The choice of pulp probably influences the qualities of the resulting paper more than any other choice. I have used abaca (Manila hemp), cotton, and linen pulps, which I buy in sheet form. Abaca is available unbleached (giving a grayish-beige color) or bleached white. The paper it produces has a kind of crispness and a hard surface. Cotton is usually bleached very white and is quite soft. Linen is the fiber I prefer. It yields a soft, crisp paper with a great deal of strength; it also takes color beautifully and tears with a feathery edge.

The first thing I do when preparing to make paper is to haul all of this stuff up from the basement. Actually, this is not too difficult a task because most of it fits into the vat for storage and moving.

After holding the sheet pulp under water, it is easily torn into smaller pieces. The pieces are then soaked in a bucket of water for a few hours or overnight. I have found that this plumps the fibers and makes them easier to separate in the blender.

After this initial soaking, I tear the pulp into smaller pieces, 1" x 1" or so, and put them into the blender with plenty of water. Too much pulp will strain the motor.

About six blenders full of pulp and enough water to come within one inch of the top of the vat makes a bath about the consistency of thin oatmeal. There should be plenty of water. I then allow this pulp bath to sit for several hours.

To form the paper I use a mold with or without a deckle. The deckle is the removable frame that creates a clean edge on the piece of paper, and holds in more pulp for thicker sheets. (When I use the mold without the deckle, I make a sheet of paper with an irregular edge. The deckle gives a more regular edge.) I dip the mold, with or without the deckle, into the vat in a scooping motion with the front edge first, and then pull up. I hold the mold above the vat tilted enough to allow the water to flow through the mesh and off the corner of the screen. I then wipe the back of the mesh with a large, dampened sponge to extract as much water as possible.

The sheet of paper is now couched (pronounced "cooched") onto the tabletop. If a deckle is being used, it is removed. The mold is placed paper-side down, and the back of the mesh is sponged to extract even more water. Then the mold is lifted, as if it were hinged to the table on one side. Lift the opposite side first and peel it off the table, thus leaving the paper behind. If I want a larger piece of paper than my mold will accommodate, I overlap the edges slightly in the couching process. When dry, the pieces will adhere to each other and make a larger sheet. (There will, however, be a slight change in the surface where the sheets overlap.) The paper should be left undisturbed to dry. Sometimes I use a fan to speed up the drying process.

After the paper is dry, I gently pick up one corner and peel it up from the table. Because it has dried on the tabletop and not on felts, it will be very smooth and flat. If I want a more irregular surface, I remove the paper while it's still slightly damp.

Usually I make 50 to 100 sheets at once. I cover the tabletop (it will hold about 24 sheets), let them dry, remove the sheets, and then repeat the process. This way I can leave the papermaking equipment out and get it all done in a relatively short time. (Otherwise, I would be spending more time setting up than making paper.)

Saddle Basket; 8"h x 9"d; photo: Rob Karosis.

Jazz Basket; 5"h x 10-1/2"d; photo: Stretch Tuemmler.

Forming the Paper

*I use a mold with
or without a deckle
to form the paper.*

*Dip the mold into the vat
in a scooping motion,
front edge first,
and then pull up.*

*Hold the mold above the vat at a
slight tilt to allow the water to flow
through the mesh and off the corner
of the screen.*

Couching

Wipe the back of the mesh with a large dampened sponge to extract as much water as possible.

If a deckle is used, it is now removed.

The paper is couched onto the tabletop.

Couching

The mold is placed paper-side down and the back of the mesh is sponged to extract more water.

The mold is lifted as if it were hinged to the table on one side, lifting the opposite side first and then peeling it off the table to leave the paper behind.

Handmade papers and leather used on the surface of the basket.

Photos: Kirby Pilcher

Coiling

There is no friendlier basketmaking technique than coiling. It requires no soaking, heavy tools, clamps, tugging, or grappling. Everything one needs fits into a bag or basket to be ready for an odd moment's work.

Coiling does have its limitations though. Coiled baskets have curves, not angles. If you want a sharp-cornered form, coiling is not for you. It is a tremendously slow process. I know of no way to hurry coiling. Obviously, the scale of materials makes a difference. The larger the core and wrapping materials, the faster the basket will develop. But with increased size comes a loss of elegance in form. Whenever I become frustrated with the time coiling requires, I think of my friends who make knotted basket forms with waxed linen, and I feel my fingers have wings by comparison.

Instructions for coiling a basket can be found in many books on basketry, including *A Modern Approach to Basketry* by Dona Meilach.

Applying the Paper Skin

This is the point at which I am always struck with a strong flash of irony. I have just spent hours and making this basket, and now I am going to cover it up. But cover it I do, and I am always pleased to look inside a finished basket to see the coiling...to see the tradition behind my own work.

To create the skin or surface, I use a soft, pliable linen paper. This is the paper I make myself. It is soft to conform to the surface of the basket, but doesn't disintegrate when wet. It has many of the same qualities as linen fabric.

The paper is torn into 1" x 1" pieces. The torn edges are feathered and soft, and blend into the adjacent pieces of paper when moistened. If the paper is cut, a subtle line will be apparent where the pieces overlap. This can be a pleasing design element if planned.

I realize that glue is definitely not within the basket artist's tradition, but it is the only way I can get the paper to adhere. I explored the possibility of using rabbit-skin glue so at least it would be natural glue, but I was dissuaded by a textile curator who said that rabbit-skin glue is terribly attractive to mice and insects and is a conservator's or collector's nightmare. I use Sobo Glue, a general craft glue that dries quickly and clear, but I think any craft glue would work.

First, a small amount of glue is spread onto an area of the basket, beginning on the bottom about three inches in circumference. The paper is then put into place petal-fashion, from the center circling out, with each piece overlapping the last and each row overlapping the previous row. Glue must not be allowed to spread onto the surface. If it does, it will create a resist and leave a white spot when the paper is painted.

When the basket is completely covered, the surface is painted with a watercolor wash with a soft, fat paintbrush. This blends the paper where it overlaps, and imparts an overall color which can be modulated to fit into the planned design. The surface is now ready for what I consider to be the real fun.

Blue Stilt Basket; 9"h x 5"d; photo: Color Marketing Concepts.

Surface Design

I think about the paper surface on the basket as a three-dimensional canvas. What happens now becomes the visual surface of the basket: the paint on the canvas.

As an undergraduate student in college, my major studies were in painting and drawing, so I tend to design for the two-dimensional surface. (Mine just happens to be wrapped around a three-dimensional object.) Actually, I do make collages which are flat and framed, as well as baskets. They share much of the same imagery.

A strong influence on the imagery for both collages and baskets comes from my textile studies in graduate school. Although each textile technique—weaving, rugmaking, crochet, knitting, batik, needlepoint, lacemaking, embroidery, netting, printing, ad infinitum—seems unrelated to the other, they all have applications in basketry. Anything is possible as long as it enhances the desired image and is structurally sound.

Usually the basis for the design on my baskets is formed with paper. I call it "cut and paste." Sometimes I use fabric as well, either linen or silk. First I assemble the papers, fabrics, and leathers I want to use. These materials have been chosen for their textures and surface qualities. Then I paint a range of colors, using watercolors, on both paper and fabric. I try to paint only as much as I will need for one basket, (pieces about 6" x 12"). I usually have some left over, but they can be saved for other projects. Because of the different textures and fibers, each paper or fabric takes the paint differently. Fabric tends to take in more paint and therefore becomes darker than paper painted in the same manner; I will dilute the color before painting the fabric if I want it to be the same as the paper. The leather takes very little color, so it must be chosen for its inherent color. This range of materials and colors becomes my palette.

Recently I have begun patterning the paper after it is painted by using colored pencils, watercolors, and pastels. It is important to me to use the best quality materials I can find, so I try to use artists' quality paints and pastels. They are more expensive, but they should last longer and be more stable. Stenciling is a particularly quick and easy way to pattern the paper. I cut patterns out of stencil paper (available from art supply stores), and either brush in powdered pastels or watercolor paints. You can cover a large area quickly and easily, controlling the color as you go. When you use this paper on the basket, you have a pattern-on-pattern quality that is reminiscent of pieced quilts.

Stamping with eraser stamps, commercial stamps, or found objects is another good way to pattern paper. It all sounds so elementary, but it can tremendously enrich the surface. Colored pencils, used randomly or in a geometric pattern, also prove to be an efficient and effective way to enhance the paper.

Fabric-weight leather adds a different texture and performs differently from paper and fabric. It is slightly stretchy and has no grain, as does fabric, and so it is particularly good for binding the neck edge of the basket. It can also be used for cut shapes, just as paper and fabric are.

Wrapped Basket; 7-1/2"h x 11"d;
photo: Craig Blouin.

Many of the designs for my baskets come from the drawing process. It is quite helpful to have some idea of where you are going when confronted with a basket surface and all that paper, fabric, and leather. You can always modify the design as you go.

After I have the drawing and have painted and patterned the materials I want to work with, I begin the actual surface design of the basket. It is a good idea to make some guideline marks on the paper skin so that you know where you are. Using a tape measure, you can mark a true line all the way around to guide you. Remember not to follow a row of coiling around, assuming that it is level, because coiling is a spiral.

Using the drawing as a guide, paper, leather, and fabric are then glued into place. Some of the patterns I make are reminiscent of pieced quilt patterns; some are woven strips of leather or fabric or simply wrapped fabric; others are almost landscape-like, a pointillist effect on paper. Let me reiterate: anything is possible as long as it enhances the image and is structurally sound.

Embellishment

Accents can be added once the surface design of the basket has been established. These are not structural elements; they don't create the surface. They are simply added on. Anything that can be stitched, tied, or glued can be used, but the integrity of materials and structure must be maintained. I would rather tie a bone onto the surface with an appropriate thread that shows than glue it. There is less stability with glue and more chance for mess. More importantly, what you see reveals the actual structure.

Sometimes what seems to be a great idea pales in the process. On several baskets I have started with one element—a leather wrapped clay marble, for instance—and planned to repeat it all the way around. By the time I got to the 50th marble and was only one-third of the way around, I began to question my design, as well as my sanity. But it is that kind of obsessive attention to detail that can make a wonderful statement.

Beads are perhaps the most obvious and easiest embellishment to attach. If you are ever in a serious bead shop, you will be impressed with the extraordinary variety of qualities characterized in beads: from the sophistication of carved ivory to the flashy arrogance of pearlescent glass to the earthy dignity of clay discs. The only problem with choosing beads is in remembering that the basket should wear the beads, not the other way around. Some beads are so dramatic they must be used carefully.

Old jewelry can provide a good source for beads. In the spirit of integrity of materials, I try to stay with natural materials, only because I think they relate more closely to my imagery.

Under the category of beads, I would include transformed objects, items which, when taken out of their normal context, can be read as beads: washers, buttons, shells, electronic components, safety pins, and so on. There is something wonderful about being lured into looking closer at a basket and discovering that those glamorous, shiny beads are indeed something so mundane as washers from the hardware store.

Another natural selection for baskets is feathers. Somehow, the nest-like quality of a basket presents a fine home for an elegant feather. I try to find a feather that has a curve or line that echoes the basket or other design elements. I don't want it to look stuck on. I find feathers frequently at the beach and in the woods; but when I'm looking for something exotic, I go to a fisherman's fly-tying shop. They often have wonderful specimens, as well as threads and fittings that are useful. As soon as you let friends know that you are interested in feathers, you will rarely open a letter without a feather falling out into your lap.

Metal can be great fun. It seems so antithetical to a lot of basketmaking materials— hard, shiny, and machine-like—but it can be used to great advantage. Wire can be used in the structure or as a design element.

Sheet metal can be fabricated into any form with modest tools: offset snips (available at hardware stores), punches, chisels, a hammer, and a file. Sheet aluminum is available from hardware stores, and sheet brass or copper from art supply stores. I cut shapes such as a bell or fish, file the edges if they are rough, and then form them. Usually these shapes are punched to accommodate a thread or leather thong for attaching; sometimes they are crimped onto a thong. There is no one right way to do it; whatever works is right.

Metal leaf is available from art supply stores, and gives an effect unlike any other. It is very thin, therefore revealing the texture beneath it when glued onto the surface. There is a traditional way to use metal leaf, but I tend to make do with acrylic medium for adhering it. It is not easy to use but is well worth the effort.

Knitting and crocheting may seem unlikely candidates for basket embellishment, and I admit I haven't used them much, but one of my favorite baskets did use a knitted linen that had been constructed on large needles to give it a lacy effect. I cut and pulled it to conform to the surface of the basket, and then stitched it into place. Because knitting is by nature stretchy, it did what I wanted it to do, and had the look of an old mariner's net encasing the basket.

Needlepoint, too, seems unlikely for basket embellishment, but stitching, much like bargello (vertical stitches), can be quite versatile and effective. The resulting design relates closely to the pattern of traditional coiled baskets. The stitches are made around each coil through the paper surface. The pattern also shows on the inside of the basket. The rug-making technique of rya knotting can be applied by knotting yarn or raffia around a coil with the pile on the outside. This gives a shaggy appearance. Embroidery stitches can enrich a surface with texture.

The simplest technique I use, and perhaps one of the most effective, is braiding or plaiting. I braid raffia and occasionally waxed linen, and glue it onto the surface of the basket to finish or mask the edges of leather, fabric, or paper shapes to keep them from looking raw. The braids look intricate and painstaking, perhaps because they are small, when actually they are so easy.

Bones, sticks, and other nonperishable objects can also be good structural or accent pieces. Sticks can become legs, handles, or latches. And bones can be buttons, latches, or design elements.

There are several ways to attach objects to a basket. The most direct method is tying. With a needle and stitching material, stitch through the basket between two coils, and then stitch back to the outside. You will have two ends to tie around or through an object. The stitching material can be thread, yarn, wire, grass, raffia, or whatever works.

Stitching is easier between rows. It is difficult to stitch through a coil. Remember, it is like stitching on fabric. You will have stitches showing on the reverse side—inside of the basket. One way to hide them is to stitch in a horizontal path so that the stitches on the inside are hidden in the creases between the coils.

Marble Basket; 12"h x 14"d;
photo: Rob Karosis.

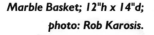

Gluing is a good way to hold things in place while you stitch or tie them on. In the long run, stitching and tying are more stable, and I think they are more in keeping with the aesthetics of handmade baskets.

In selecting objects for my baskets, I try to keep in mind the basic design concepts of color, line, and proportion. I look for relationships that are already there, and for ways the basket and the object can enhance each other.

Surface Finish

Baskets are friendly, intimate objects. They should be touched, held, and used. But if they incorporate handmade paper, there can be a problem. Paper, in its natural state, is not very resistant to abrasion and soiling. In an effort to deal with that problem, I discovered acrylic medium. (I had used it when painting with acrylic paints, and so had some idea of its properties.) It is a white liquid which dries clear; it can be diluted with water, but is water-proof when it dries. It can be tinted with any water-based pigment, and can be either glossy or matte.

By using a combination of gloss and matte medium, I can control how much sheen the surface will have. I use washes of the medium (about one-third medium, two-thirds water) tinted with watercolor paints to build up a series of coats. One coat of full-strength acrylic medium makes the basket look as if it has been shrink-wrapped. By building up five or six coats of diluted medium, the effect is softer and more subtle, and the color can be more modulated. Allow each coat to dry before adding the next. The last coat should be two-thirds medium, one-third water. The surface should be completely covered. I paint everything except feathers with the acrylic medium. This gives a cohesive quality to the surface.

The final surface treatment is usually a leather finish, not unlike shoe polish, which comes in various shades of brown, tan, and mahogany. Use it sparingly to buff the surface. It settles in the low places, enriching the surface. Sometimes I don't use it at all, but it can add a lovely extra dimension to the basket.

The relationship of paper to basket is what the artist makes of it. For me, it is a way of relating two ways of seeing: two-dimensional and three-dimensional. It is a way of using materials I like to create forms I like. It is a way of incorporating color into basketry and of creating a surface for embellishment.

For you, using paper and basketry together may open up new possibilities for extending the tradition and for making your own personal statement. I hope so.

Autumn Wind Basket;
17"h x 12"d;
photo: Craig Blouin.

Sky Sails I; 1986; coiled in waxed linen with ash, driftwood, paint, and paper; 4" x 5-3/4" x 2-1/2".

GAMMY MILLER

My work is about that which is evanescent or endangered. I borrow freely from the detritus of my life, and other forms of life, for materials and inspiration.

Several years ago I began looking for new materials to add to my repertoire of ash, mud, driftwood, and found objects, and fell upon paper. With this paper, made from recycled mail and plant fibers (especially birch), I have been able to extend and change the rather static forms I previously had been using. Paper led to accretions of paper, to paint, and finally to the written word.

On this lovely and rather crude paper, I leave messages—sometimes concealed, sometimes bitten by fire—in each of my works because I can no longer live and work in silence.

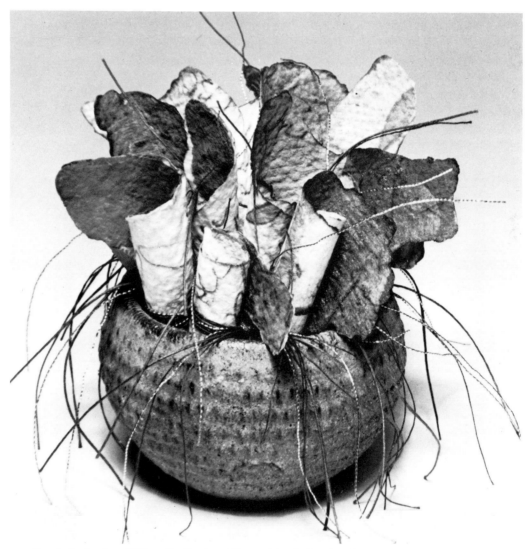

Scrolls for Justice; 1986; coiled in waxed linen with ash, handmade paper, and paint; 2-1/2" x 3".

Untitled (kite) I; coiled in waxed linen with mud, driftwood, paint, and paper.

Ash Basket with Pierced Scrolls II; coiled in waxed linen with ash, driftwood, and handmade paper.

Sky Sails II; 1986; coiled in waxed linen with ash, paper, and paint; 4" x 5-1/4" x 2-1/2".

Callahuayas Book; 1984; coiled in waxed linen with mud, paper, and talismans; 3-1/4" x 3".

Justice Scrolls II; 1986; coiled in waxed linen with mud, paint, and paper; 2-1/2" x 3".

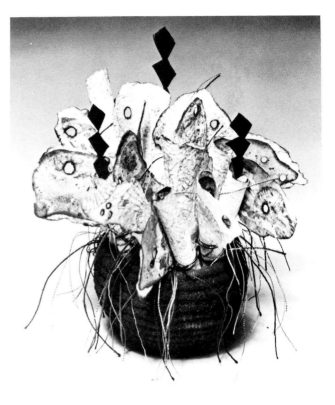

Spring Rolls; 1986; coiled in waxed linen with mud, paper, and paint; 3-1/4" x 3-1/2".

Untitled (kite) II; 1986; coiled in waxed linen with ash, driftwood, paper, paint, and potato bug; 3-1/2" x 4-3/4" x 2-1/4".

Back view of Untitled (kite) II.

Walls of Acoma; handmade abaca paper and dyed linen.

I'm Not Fooling This Time; bleached linen and handmade cotton paper; painted.

ELLEN CLAGUE

My basketry work with handmade paper is concerned with a paper skin applied to a coiled linen basket. The process is simplicity itself, once you know how it's done. My system involves the application of wet pulp, as opposed to a dry-sheet method. Once the coiled basket is completed, the wet pulp is applied, dried, and often painted and decorated. Within the basic premise of a skin over a core basket, my choices of materials or methods are based on structural or aesthetic considerations. How the basket is coiled, the paper applied, and which pulp I use are the variables that give each basket its character.

The first choice is the linen basket itself. I often used hand-dyed and hand-spun wools to coil baskets, but once I began wanting to apply the color directly onto the basket, linen seemed the appropriate choice. I had painted linen canvas in traditional painter's fashion, and linen warps for painted warp techniques. Linen has a wonderful sheen, is easy to work with, and achieves the desired look when painted. It is readily available in one-half or one pound cones from weaving supply sources. I use both natural and bleached linens. While I prefer the smooth long staple, tightly spun linens, I have used some fuzzier ones which give a nice surface. (I was breathing fuzz for weeks while I worked, though, and that concerned me.) Choosing between bleached white and unbleached tan is simply a question of aesthetics, as is the using of a single thread for a neater surface or a double thread for a faster finishing but less controlled surface.

The next choice is the stitch. I coil in either the figure-eight or lazy stitch. There is a structural choice and sometimes a time consideration. Figure-eight takes *forever* but gives the strongest basket for direct application of wet pulp. The figure-eight is a wrap around the coil and a stitch into the row below each time. The lazy stitch allows the coil to be wrapped several (three to five) times and then a stitch into the row below. This creates a slightly irregular surface, which I have never found to be a problem with as fine a thread as the linen. The main difference is that a basket coiled in lazy stitch has less stuctural rigidity.

For a basket on which I intend to form a thick skin with a controlled surface, I choose figure-eight, apply my pulp directly, and form it on the basket. A thinner skinned basket can be done in lazy stitch. I then sheet-form the paper and leaf-cast the wet pressed sheets. This puts less stress on the basket form, allowing the less rigid, less time-consuming stitch to be used.

My core material is either Fibre-Flex or a cellulose upholstery welting cord. For linen baskets, I find the welting cord compatible in size. It is available at upholstery shops on large rolls.

Prepared Basket

A small basket has been readied, using upholstery welting cord and 8/2's bleached linen.

The basket is ready to be covered in paper. The basket has been coiled in a lazy stitch.

After completing a coiled basket—a three-dimensional canvas, in effect—my next choice involves the paper itself. While I have made delightful papers from day lilies, corn husks, etc., which would be wonderful for baskets, they don't lend themselves to the sort of painted surface I'm doing. So I choose abaca (Manila hemp), cotton linter, sisal, or flax pulps, which I buy in a compressed sheet from a papermaker's supply house. These sheets are ready-beaten fibers and simply need re-hydrating in small amounts in a blender full of water. Which pulp I choose is an aesthetic decision based on color, surface texture, and softness or resilience. Since I leave the insides of the baskets uncovered and unpainted, I like to use the linen. Its color and hardness or fuzziness blends with the papers. A fuzzier, unbleached linen works well with sisal or flax. When I made the baskets in the fool's porcelain pot series, where the surfaces resembled pottery, I used cotton linter on a figure-eight stitched basket of bleached linen.

Forming Paper

Using a small 5" x 7" mold and deckle, a sheet of linen paper is pulled from a vat of water and linen pulp.

After pulling the sheet of paper, excess water must be drained before the sheet has enough structural integrity to be couched. The process of unmolding onto a stack of felts is called couching.

Forming Paper

Once excess water has drained, the mold can be inverted and the paper will remain intact. A board and several dampened felts are stacked underneath the first sheet of paper.

Pressing Paper

Pressing one edge of the mold on the felt stack, and using a rolling motion, the paper is transferred from the mold to the felt.

After several sheets have been formed and layered on the stack of felts, more felts and another board are added to the top, and the post is placed in the press.

Excess water is squeezed from the paper. I favor a small antique book press for the size sheets I am making here.

Using Flexible Screen

*The post is removed from the press. The papers can now be
removed from the felts. Paper not being applied to a basket
would be placed on a drying surface at this point.*

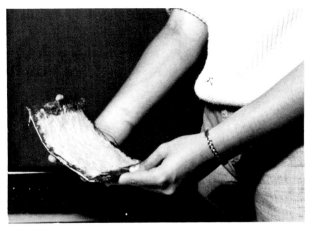

*Paper can also be pulled from the vat with a flexible
screen made of two layers of non-rust window screening
that have been bound around the edges with duct tape.
This is trickier than using the mold and deckle because the
deckle functions as a fence to keep the slippery wet pulp
on the mold screen.*

*The advantage of the flexible screen is that it can be curved in
your hand and shaped to fit the basket.*

After re-hydrating, the pulp is put in a vat of water. Nine parts water to one part pulp is standard, but I use much less water when casting thick sheets or using a flexible screen. Sheets are then pulled with a mold and deckle or a flexible screen. For the porcelain series, I used a flexible screen to sheet form. The screen consists of two layers of non-rust window screening bound around the edges with duct tape. By not using the conventional paper mold, I can then cast directly onto the basket. Once a couple of sheets are cast, I begin pressing out the water and bonding the sheets together. This is a very tedious process, but I like the results. I hold a large sponge inside with one hand, and another on the outside with the other hand. By pushing with both hands, the water is pressed out and bonds are formed between the sheets and the basket without distorting the shape. I use a piece of screening to protect the paper from too much patterning from the sponge, and it also reduces pull-off problems. Layer after layer is applied in this way, from the bottom up and around the basket, until a thick, rich skin surface is achieved.

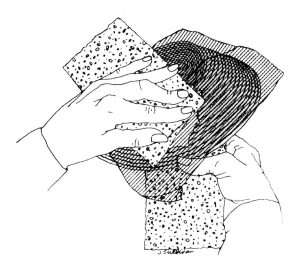

Instead of removing the paper sheet from the screen onto a felt, I couch directly onto the basket itself. Paper made directly onto the basket seems to be more integral.

Water is pressed out using sponges. A small piece of fiberglass screening is useful to keep the paper from coming away with the sponge. I also like the texture the screening imparts.

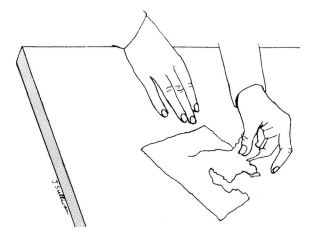

Although still quite wet, the paper has considerable strength and is quite easy to work with.

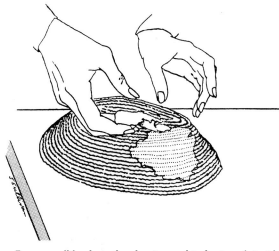

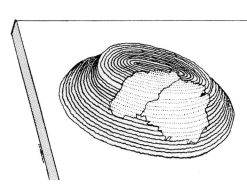

For a small basket, the sheets need to be torn into pieces to mold around the curves. Several pieces at a time are placed on the basket.

Methyl cellulose (wallpaper paste) is useful in several ways: coating the basket before applying a piece of paper, bonding the edges of two papers, and as a protective coating on the finished product.

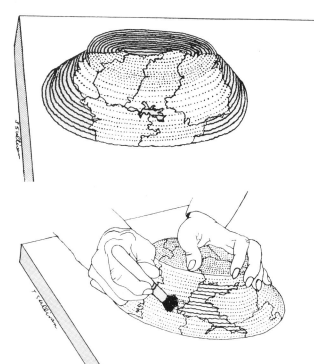

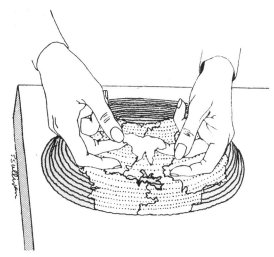

Using a painter's stippling brush and paying particular attention to the edges to be joined, I press the paper into the basket surface.

Now covered in paper, the basket is dried in the oven. This is necessitated by the humid climate where I lived.

Let me put in a plug here for methyl cellulose, a type of wallpaper paste available from papermakers' suppliers or wallcovering stores. Methyl cellulose is great stuff: it is archival, dries clear, and works like a charm. My rule of thumb is that I add it when I think of it. It can be an additive in the vat of pulp, brushed on the basket surface just before building up with more layers, or even brushed on a dried basket as a fixative. It could be used in all three stages to create a stiff basket.

Methyl cellulose is equally useful in the alternative method of pulling and pressing sheets before applying them to the basket. In this method, I use a small conventional mold and deckle and pull 3" x 5" sheets, which are couched onto felts and pressed in a small book press. The sheets are then laid onto the basket. Generally I rip them into even smaller pieces as I am working. I then do a process known as leaf casting, which uses a stipple brush or sponge to blend the edges together and bond the paper to the basket contours. Gently press a few pieces at a time, and repeat until the basket is covered. Depending on your pressing, pulp, or other variables, additional water may need to be pressed out with the sponge. If the pressed paper is too dry, mist the edges with a sprayer or apply methyl cellulose to form an adequate bond.

Now I have a wet object, in fact an extremely wet one if direct casting has been done. I do more papermaking in warm weather months, and in my humid Connecticut woods I am concerned about mold and mildew formation in things that take weeks to dry. While high heat is not good for paper, mold is a worse enemy so I dry the basket in the oven. I use a fine wire baking rack under the basket, and place it in a preheated oven. I start at a setting of 250 degrees when the basket is wet, checking it every 10 minutes and gradually lowering the temperature. As it nears drying, I reheat the oven, turn it off, and put the basket in for half an hour. I repeat as needed, taking the basket out each time I turn on the oven. I usually air dry for several more days before painting on the paper.

The paper shrinks as it dries, accentuating the coiled surface underneath. I recently had the opportunity to work with linen pulp, which has an extreme rate of shrinkage with interesting possibilities. The shrinkage of the chosen pulp is yet another aesthetic variable. Please keep in mind that if you place the paper inside the basket, it shrinks *AWAY* from the basket and becomes a free-standing paper form.

My final step is painting and sealing the basket. I use regular artists' tubes of acrylic paints and gel medium, and often the pearlescent pigments, for a glorious sheen which is softly echoed by the sheen of the linen inside the basket. Pastels, watercolors, or colored pencils are all appropriate choices for applying the decorative touches. As I mentioned, methyl cellulose is a good final coating, as are any of the artists' fixatives. Paper needs protection, and baskets usually sit in the open and get handled at shows.

One last variable is powdered pearlescent pigments in the pulp. Many of my baskets are painted with metallic pigments, so I like to use a metallic in the pulp itself. I realize that paper may not show after painting, and so they may not have any effect, but it seems fitting so I do it. After all, I just covered up a basket that I spent months coiling. I'm beyond questions at this point. The pigments, the necessary retention aids, and directions are available from papermakers' supply houses. They are powders, so masks, gloves, and care should be used.

I'm never totally sure where a choice may lead until I try it out, and I certainly don't know what choices and experiments I may yet make in my work. The most exciting of all to me are the choices that you may make, using my techniques in your own work. Have fun!

Acrylic paints mixed with gloss medium are layered on the surface. I don't paint the insides of the bleached linen baskets because they are beautiful in their own right.

Once the basket has thoroughly dried (usually a wait of several days), I use a scalpel to trim just inside the lip of the basket.

Silk threads, which hold an agate in place in the interior, have been stitched around one side of the completed basket. They have been painted and drawn with pastels. At this point, the basket has a personality to me and has been named **Heartplace.**

Darkroom technician - Patricia Andriese

Ya-Papa; 1987; 19" x 22" x 3".

Master of Women, Innocence, Truth, Devotion; 1987; 18" x 22" x 3".

KARRON NOTTINGHAM HALVERSON

In my work in handmade paper and mixed media constructions, I attempt to redefine the surface patterning process by means of decontextualization...that is, combining materials and images, however out of context they appear to be. The manipulation of a variety of materials to create symbolic images and environments is the key to the language of my forms. Nameless animals and various other objects find space among layers of color relations. Applied portions are used to define space, as color defines environment.

The presence of the *Dragon-Dog/Cat* image in my work is a means of expressing satirical life plays on the essence of womanhood. These fragmented tile constructions contrast more traditional treatments of theme...

My work in handmade paper and mixed media constructions is primarily a self-motivating adventure in color...emphasizing texture and shape relationships. The handmade paper forms of cotton and abaca fibers are pulped, utilizing the Western method of papermaking. Acrylic powders are added to the pulp for durability and weatherproofing measures. The pulp is molded by hand into desired shapes one-half inch thick, pressed, and left to dry. These forms are then treated with a compound, not to alter the paper's surface texture, but to prepare the surface for greater saturation of color. Each individual unit is then finished, using a variety of colorants: pastels, crayon, acrylic, colored pencil, etc. Then the units are assembled in layers to created the intended image. An acrylic sealer is used as a final preservation method.

Stick it Where?; 1987; 19" x 22" x 3".

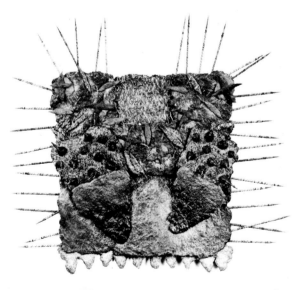

Splendor Active Principle; 1987; 23-1/2" x 25-1/2" x 5".

Golden Bowls; sewn tea chest paper and obituaries; collection of B. Saniie.

DONNA RHAE MARDER

Coffee Cups; sewn coffee filters; photo: M. Tropea.

My first pieces using the technique of sewn paper related directly to the reading I had been doing about traditional quilts and piecing techniques. Those first works were primarily decorative and concerned with rearranging and altering flat pieces of paper. For example, fragmenting and shifting a map of Scandinavia and an icy-looking rice paper; appliquéing a map of Hawaii in the style of an Hawaiian quilt; adding a third dimension to a traditional pattern executed in papers of wood veneer.

Later pieces are more sculptural and personal in content. Sometimes I incorporate tiny books and envelopes in quilt patterns. Most often, I relate specific papers and shapes. Letters turn into gloves, suggesting hands. Recipes transform into aprons, reminiscent of wives and mothers. Bowls are made of world maps or obituaries. Cups are sewn from coffee filters or photocopies of earlier cups. "Junk" reading becomes a bathrobe.

Generally, I do some thinking and drawing before I start a piece. But the process of cutting, piecing, and sewing, the nature of the paper, and the limits of how I can construct paper objects on my sewing machine always influences the finished piece. Fairly often, what I create is not what I anticipated.

I am aware of conservational issues about my work. I de-acidify papers that are unstable, and advise purchasers against placing works in direct sunlight. I also make the pieces as sturdy as possible, given the limits of the material. I consider the fragile transitory aspect of my work a conscious reflection of my understanding that that's the way life is.

■ Bibliography

Ballinger, Raymond A. *Design with Paper in Art and Graphic Design*. New York: Van Nordstrand Reinhold Co.

Barrett, Timothy. *Japanese Papermaking*. NY: 1984.

Bell, Lillian A. *Plant Fibers for Paper Making*. McMinnville, OR: Liliaceal Press, 1984.

.*Papyrus, Tapa, Amate & Rice Paper.* McMinnville, OR: 1985.

Boertzel, Barbara Moon. "Papermaking in Micronesia." Grant Project, Agana, Guam.

Farmer, Jane, ed. *Paper as Medium*. The Smithsonian Institution.

Cunning, Sheril. *Handmade Paper*. Escondido, CA: Hatrack Press, 1983.

Grummer, Arnold. *Paper by Kids*. Minneapolis, MN: Dillon Press, 1980.

Heller, Jules. *Papermaking*. New York City: Watson-Guptill, 1978.

Hughes, Sukey. *Washi: The World of Japanese Paper*. New York: Kodansha International Ltd.

Hunter, Dard. *Papermaking: The History and Technique of an Ancient Craft*.
    New York: Dover Books, 1978.

Johnson, Pauline. *Creating With Paper*. Seattle: University of Washington Press.

Koretsky, Elaine. *Color for Hand Papermakers*. Brookline, MA: Carriage House Press, 1983.

Krohen, Val. *Hawaii Dye Plants and Dye Recipes*. Hawaii.

Mason, John. *Paper Making as an Artistic Craft*. Faber Paperbacks.

Norita, Kiyofusa. *A Life of Ts'Ai Lung and Japanese Papermaking*. The Paper Museum, Tokyo.

Sargeant, Peter T. *Hand Paper Making Manual*. Paper Make Covington, VA.

Studley, Vance. *The Art and Craft of Handmade Paper*. New York: Van Nostrand Reinhold Company, 1977.

"Tapa, Washi and Western Handmade Paper". Honolulu Academy of Arts, 1980.

Toale, Bernard. *The Art of Papermaking*. Worchester, MA: Davis Publications, Inc., 1983.

VonHagen, Victor Wolfgang. *The Aztec and Maya Papermakers*. New York: Hacker Act Books, 1978.

■ Papermaking Suppliers

Asao Shimura
Cannabis Press
431 Fukuhara Kasama-shi
Ibarake-ken 309-15 Japan

Buckeye Cellulose Corporation
P.O. Box 8407
Memphis, TN 38108

Carriage House Handmade
　Paper Works
8 Evans Road
Brookline, MA 02146
Send $1.00 for catalog

Cerulean Blue, Ltd.
P.O. Box 21168
Seattle, WA 98111-3168
Send $3.00 for catalog

Cheney Pulp and Paper Company
P.O. Box 60
Franklin, OH 45005

Daniel Smith Inc.
4130 1st Avenue S.
Seattle, WA 98134

Gold's Art Works Inc.
2100 N. Pine Street
Lumberton, NC 28358

Kensington Paper Mill
2527 Magnolia Street
Oakland, CA 94607

Lee S. McDonald
P.O. Box 264
Charlestown, MA 02129
Send $3.00 for catalog

Nova Color
5894 Blackwelder Street
Culver City, CA 90230

Paper Source
1506 W. 12th Street
Los Angeles, CA 90015

Straw Into Gold
3006 San Pablo Avenue
Berkeley, CA 94702

Twinrocker Papermaking
　Supplies
RFD 2
Brookston, IN 47923

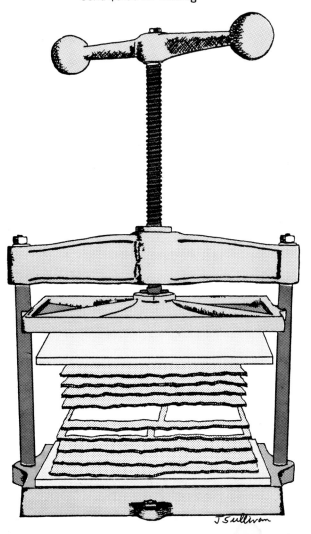

A book press showing the layers of board, felts, paper, felts, and another board. Illustration: Jane Sullivan.

■ Glossary

Pam Barton's wood tapa beaters and tapa-making anvil are used for breaking down plant fibers. Photo: Dina Kageler, Volcano, HI.

abaca (Musa textalis) **-** This is the Philippine word for Manila hemp, the fiber that comes from the stalk of a special type of banana tree.

archival adhesive - Any adhesive that does not discolor, become brittle, lose its adhesion or cause deterioration of the paper with age.

archival paper - Paper with a pH of 7.0, which means that it has good aging properties.

bast fiber - The phloem, or inner bark fibers, of woody plants.

beater - Any piece of equipment used in breaking down the fiber for papermaking. For example: a blender, heavy-duty mixer, mallet, or anything you can devise to pound the fibers with.

bonding - The ability of the fibers to adhere to one another.

buffering agents - Substances which provide an alkaline reserve to counteract the effects of acidic chemicals and pollutants in the atmosphere. Calcium carbonate and magnesium carbonate are two examples.

caustic soda - Sodium hydroxide or lye used in cooking fibers.

calcium carbonate - A buffering agent that is added to pulp to protect the resulting paper from future acidity that may be picked up from the atmosphere.

cellulose fiber - Fibers derived from plants.

coiling - A simple stitched construction where a thin pliable material is used to stitch concentric circles over a single material or group of materials.

cotton linters - The seed hair of the cotton seed after the staple cotton has been removed; widely used for papermaking.

couching - The process of transferring a newly formed sheet of paper from the screen to the drying surface.

dispersing agent - An ingredient added to the pulp that allows each fiber to be independent and not cling together, making a smoother, more even sheet of paper.

deckle - The open wooden frame that fits on top of the mold. It contains the fiber and defines the edge of the sheet.

direct dye - A class of dyestuff for use on cotton, linen, and viscose rayon.

Fiber-Flex - A paper cord with netting over it. Usually available from basketry supply stores. Used primarily for coiled baskets.

fiber-reactive dye - A class of dye designed specifically for cool water dyeing. It is mixed with salt and washing soda for use on cotton, linen, and silk.

fibrillate - To flatten and split the fiber. This causes fibrils (smaller fibers) on the surface of the fiber, and is the result of beating the fibers.

figure-eight stitch - A stitch used in coiling that produces a very strong basket.

formation aid - Used in papermaking as a deflocculent for long-fibered pulps. It keeps the fibers from entangling.

hardware cloth - A galvanized wire screen with a half-inch mesh.

Hollander Beater - A special beater designed for processing large amounts of pulp.

hydraulic press - Large press for extracting moisture from sheets of paper.

hydrogen peroxide - A solution used as a bleaching or disinfecting agent.

keta - Hinged wooden frame which holds su; used as a papermaking mold.

kraft paper - Strong wrapping paper, usually brown, made from sulfate pulp.

lazy stitch - A stitch used in coiling; not as structurally sound as the figure-eight stitch.

lignin - An organic substance closely allied to cellulose which forms the essential part of woody fiber.

linen - A natural fiber produced from the stem of the flax plant.

linters - A general term for pre-processed pulp. This can be purchased in either sheet or chip form.

lye - See caustic soda.

methyl cellulose - Available from papermaking suppliers, this can be used as sizing or as an adhesive, or internally as a binder and strengthener.

mold - Any frame covered with a screen or surface on which a sheet of paper can be formed.

mordant - A material which fixes dye onto a fiber; usually called for when using vegetable dyes. Salt and vinegar are some mordants in chemical dyeing.

natural dyes - Dyes made from materials from nature.

neri - The general term used for formation aid or dispersing agent. Usually made from vegetable mucilage. In Japan, neri comes from the tororo-aoi (Hibiscus manihot) root.

olla - A large-mouthed jar or pot of earthenware.

papier-mâché - A material made of paper pulp mixed with rosin, oil, etc., that can be molded into various objects while moist.

papermaker's felt - Any material placed between the wet sheets of paper during the couching process. Examples: Pellon, sheets, blankets, printmaking felt, or heavy blotter paper.

Pellon - A matted material used as a lining.

phloem - Bast or inner fibers of woody plant.

PNS - A dispersing agent.

post - A stack of felts and couched sheets of paper.

Procion - Trade name of a fiber-reactive dye.

pulp - The resultant of beaten fibers.

raffia (Malagasy rafia) **-** The fiber of the raffia palm; used for tying plants and making baskets and hats.

retting - The breaking down of fiber by the natural rotting process. This process can be used in lieu of cooking.

second cut cotton linter - The second cut seed hair of the cotton seed after the staple cotton has been removed; it is a short fiber, the least expensive, and the easiest pulp with which to work.

sisal - A strong fiber obtained from the leaves of an agave.

slurry - The term used for the pulp and water combination in the vat.

sizing - A material added to the vat or finished paper that decreases the absorbency of the paper.

Sobo Glue - Brand name of a general craft glue that remains flexible when dry.

sodium carbonate - Soda ash, washing soda; the most commonly used alkali for cooking fibers to prepare them for beating. It dissolves impurities in the pulp.

su - The flexible bamboo screen used in Oriental papermaking.

tapa - The generic term in Polynesia for bark cloth made by beating the inner bark of specific plants and trees to a fabric-like material which, during ancient times, served every purpose that cloth serves us today.

tororo-aoi - Hibiscus manihot, sometimes known as edible hibiscus. The root of this plant is crushed and added to water to produce the forming aid used in Oriental papermaking.

TSP - Trisodium phosphate; a mild alkali which can be substituted for washing soda.

twining - A weaving process in which two or more weavers are twisted around one another as they interlace with the warp.

warp - The elements on which weavers will interlace at right angles to produce a surface.

water leaf - Unsized paper.

watermark - A design or mark that is put onto the mold and produces a mark on the finished sheet of paper.

weaver - The elements crossing the width of the warp.

yin/yang - Chinese dualistic philosophy in which yin, the passive female cosmic element, is opposite but always complementary to yang, the active masculine element.

■ Index

A
Abaca, 9, 126, 160
Acrylic paints, 67
Adhesives, archival, 160
Archival paper, 160

B
Barton, Pam, 53-62
Bast fibers, 10, 160
Beaters, 160
Bock, Sharon, 109-119
Bonding, 160
Buffering agents, 16, 160

C
Casting, 53-62, 71
Caustic soda, 160
Cellulose, 10, 160
Clague, Ellen, 144-151
Coiling, 160
Color, 67, 110
Couching, 22, 111, 131, 149, 160

D
Dahl, Carolyn, 80-82, 89, 90
Deckle, 18, 20, 160
Dispersing agent, 160
Drying, 22, 41, 95
Dyes, 16, 67, 110, 160

E
Equipment, basic, 11, 12, 130
Embellishing, 67, 71, 137, 138

F
Felt, for papermaking, 161
Fiber-Flex, 160
Fibrillate, 160
Finishes, 139
Formation aids, 160
Fulkerson, Mary Lee, 64-69

G
Glossary, 160, 161
Gonzalez, Rosemary, 70-74
Grass fibers, 10

H
Halverson, Karron Nottinham, 153, 154
Hardware cloth, 160
Hunter, Lissa, 125-139
Hydraulic press, 160
Hydrogen peroxide, 161

I
Introduction, 4

K
Keta, 161
Kraft paper, 161

L
Layered pulp, 48
Leaf fibers, 10
Lignin, 161
Linen, 161
Linters, cotton, 160, 161

M
Marder, Donna Rhae, 155
Materials
    Plant fibers, 9
    Leftover basketry, 9

Merkel-Hess, Mary, 120-124
Methyl cellulose, 26, 161
Miller, Gammy, 141-143
Molds
How to use, 18, 19, 71, 92
    Making your own, 18, 19, 71, 72, 92

Mordants, 14, 161
Mulford, Judy, 41-45

P
Papier Mache, 66, 161
Phloem, 161
Plant fibers, recommended
    varieties, 9, 10, 32-39, 48, 92, 105-108
Plaster molds, 71, 72

Plaster of Paris, 71
PNS, 161
Posts, 161
Precautions, safety, 12
Pulp, definition of, 161

R
Raffia, 161
Removal, of sheets, 97
Retting, 161
Rolled paper, 97

S
Salmont, Beth, 75-78
Seed fibers, 10
Sewn paper, 155
Sisal, 161
Sizing, 16
Slurry, 20, 161
Smith, Sue, 9-39
Sodium carbonate, 11, 161
Spun paper, see rolled paper,
Storing fiber, 94
Suspension, of fibers, 20

T
Tapa, 161
Tips, general, 4, 5
Tools, 11, 12
Trisodium carbonate, 11
Trisodium phosphate, 161

V
Vats, 96, 110, 129

W
Wand, Alice, 48-51
Washing fiber, 94
Water leaf, 161
Watermark, 161
Waterproofing, 111
Weaving baskets with paper, 52-62
Wold, Marilyn, 91-108